Sabrina Schulze

Structural basis of the L-carnitine/γ-butyrobetaine transport in CaiT

Sabrina Schulze

Structural basis of the L-carnitine/γ-butyrobetaine transport in CaiT

Südwestdeutscher Verlag für
Hochschulschriften

Imprint

Any brand names and product names mentioned in this book are subject to trademark, brand or patent protection and are trademarks or registered trademarks of their respective holders. The use of brand names, product names, common names, trade names, product descriptions etc. even without a particular marking in this work is in no way to be construed to mean that such names may be regarded as unrestricted in respect of trademark and brand protection legislation and could thus be used by anyone.

Cover image: www.ingimage.com

Publisher:
Südwestdeutscher Verlag für Hochschulschriften
is a trademark of
Dodo Books Indian Ocean Ltd., member of the OmniScriptum S.R.L Publishing group
str. A.Russo 15, of. 61, Chisinau-2068, Republic of Moldova Europe
Printed at: see last page
ISBN: 978-3-8381-2696-8

Zugl. / Approved by: Frankfurt: Johann-Wolfgang Goethe University, University, Diss., 2011

Copyright © Sabrina Schulze
Copyright © 2011 Dodo Books Indian Ocean Ltd., member of the OmniScriptum S.R.L Publishing group

Structural basis of Na^+-independent and cooperative transport in the L-carnitine/γ-butyrobetaine antiporter CaiT

Dissertation
zur Erlangung des Doktorgrades
der Naturwissenschaften

vorgelegt beim Fachbereich Biochemie, Chemie und Pharmazie
der Johann Wolfgang Goethe – Universität
in Frankfurt am Main

von
Sabrina Schulze
aus Forst/Lausitz

Frankfurt am Main, 2010
(D30)

Die Arbeit wurde in der Abteilung Strukturbiologie des
Max-Planck-Instituts für Biophysik in Frankfurt am Main durchgeführt
und vom Fachbereich Biochemie, Chemie und Pharmazie der
Johann Wolfgang Goethe – Universität als Dissertation angenommen.

Dekan : Prof. Dr. Dieter Steinhilber
1. Gutachter : Prof. Dr. V. Dötsch
2. Gutachter : Prof. Dr. W. Kühlbrandt

Datum der Disputation : 18.01.2011

To Beatrice S. M. Buchin

Zusammenfassung

Der L-Carnitin/γ-Butyrobetain Antiporter CaiT ist ein Mitglied der Betain/Carnitin/Cholin Transporter (BCCT) Familie. Sekundärtransporter der BCCT Familie transportieren Substrate, die eine positiv-geladene quartäre Ammoniumgruppe besitzen. CaiT besteht aus 504 Amiosäuren und besitzt ein moleculares Gewicht von etwa 56 kDa. In Enterobakterien wie *Escherichia coli*, *Proteus mirabilis* und *Salmonella typhimurium* wird die Expression des *caiTABCDE* Operons unter anaeroben Bedingungen induziert. Unter diesen Bedinungen ist CaiT der Haupttransporter des Betain-Derivates L-Carnitin. In Enterobakterien wird L-Carnitin unter anaeroben Bedingungen aufgenommen und dehydratisiert wobei Crotonobetain ensteht. Crotonobetain wird anschließend zum Endprodukt γ-Butyrobetain reduziert. γ-Butyrobetain ist das Gegensubstrat, das aus der Zelle hinaustransportiert wird, wenn L-Carnitin in die Zelle aufgenommen wird. Der Austauschmechanismus von L-Carnitin gegen γ-Butyrobetain geschieht ohne das Vorhandensein eines elektrochemischen Gradients, d.h. CaiT ist sowohl H^+- als auch Na^+-unabhängig.

Ein Ziel dieser Arbeit war es die drei-dimensionale (3D) Struktur von CaiT mittels Röntgenstrukturanalyse zu lösen. Weiterhin sollten mit Hilfe der 3D-Struktur und funktionellen Studien detailiertere Erkenntnisse über den kationenunabhängigen Antiportmechanismus von CaiT ermittelt werden.

Im Rahmen dieser Arbeit wurden die 3D-Röntgenkristallstrukturen von drei CaiT-Homologen der Enterobakterien *P. mirabilis* (PmCaiT), *E. coli* (EcCaiT) und *S. typhimurium* (StCaiT) mittels molekularem Ersatz (engl.: *molecular replacement*, MR) mit einem Alanin-Model des CaiT verwandten Na^+/Glycinbetain Symporters BetP gelöst. PmCaiT konnte mit einer Auflösung von 2.3 Å gelöst werden. Das Protein kristallisierte in der Kristallraumgruppe *H*3, mit drei Molekülen in der asymmetrischen Einheit (engl.: *asymmetric unit*, AU). Die drei PmCaiT-Moleküle ordneten sich innerhalb der AU um eine kristallographische dreifach Symmetrieachse

I

an. EcCaiT wurde mittels MR mit einem Alanin-Model von PmCaiT bei einer Auflösung von 3.5 Å gelöst. EcCaiT kristallisierte in der Kristallraumgruppe $P3_2$, ebenfalls mit drei Molekülen in der AU, jedoch ohne kristallographische Symmetry. Während der Verfeinerung des EcCaiT-Models wurde eine strenge dreifache nicht-kristallographische Symmetry (engl.: *non-crystallographic symmetry*, NCS) angewandt. StCaiT, das ebenfalls mittels MR mit einem Alanin-Model von PmCaiT, jedoch bei einer Auflösung von 4.0 Å gelöst wurde, kristallisierte in der Kristallraumgruppe $P6_5$, ebenfalls mit drei StCaiT-Molekülen in der AU, ohne kristallographische Symmetry. Bei der Verfeinerung des StCaiT-Modells wurde wie bei EcCaiT eine strenge NCS angewandt.

Da die Auflösung von 4.0 Å bei StCaiT zu niedrig ist um detaillierte moleculare Erkenntnisse zu gewinnen, wurden Protein- sowie Substratinteraktionen nur an den Strukturen von PmCaiT und EcCaiT analysiert. Alle drei CaiT-Homologe weisen jedoch einen ähnlichen strukturellen Aufbau auf.

In der Röntgenkristallstruktur bildet CaiT ein symmetrisches Trimer, das über ionische und polare Wechselwirkungen zwischen den Protomeren stabilisiert wird. Der trimere Oligomerisierungszustand von CaiT in Detergenzlösung sowie in zwei-dimensionalen Lipidmembrankristallen wurde bereits in früheren Arbeiten gezeigt. Jedes der drei CaiT-Protomere besteht aus zwölf Transmembranhelices (TMH), die N- und C-terminalen Domänen des Proteins befinden sich auf der cytoplasmatischen Seite. Zehn der TMH bilden zwei invertierte Wiederholungseinheiten aus jeweils fünf TMH. Die erste Einheit besteht aus den TMH 3 – 7, die invertierte zweite Einheit besteht aus den TMH 8 – 12. Beide Wiederholungseinheiten sind strukturell nahezu identisch und lassen sich fast vollständig übereinanderlegen, jedoch weisen die Aminosäuren der beiden Einheiten keine signifikante Sequenzidentität auf. Die ersten beiden Helices der Wiederholungseinheiten, die TMH 3 – 4 und die TMH 8 – 9, bilden ein antiparalleles vier-Helix-Bündel, in dem in CaiT zwei Substratbindestellen lokalisiert sind. Eine derartige Transporterarchitektur wurde erstmals in der Struktur des Na^+/Alanin Symporters $LeuT_{Aa}$ des thermophilen Bakteriums *Aquifex aeolicus* gezeigt. Bislang wurden, inklusive CaiT, sieben Sekundärtransporterstrukturen gelöst, die diese LeuT-Transporterarchitektur aufweisen. Ungewöhnlich dabei ist, dass diese sieben Sekundärtransporter fünf verschiedenen Transporterfamilien angehören und eine Verwandschaft auf Basis der Aminosäuren nicht zu finden ist. Da jedoch die tertiäre Struktur dieser Tansporter konserviert ist, kann davon ausgegangen werden, dass sie alle von einem Urprotein entstanden sind, welches zunächst aus fünf TMH bestanden haben muss. Im Laufe der Evolution hat sich das Urgen des Urproteins zunächst dupliziert und die weitere Evolution hat zwar die Aminosäuresequenz verändert und den Umweltbedingungen angepasst, jedoch ist die tertiäre Struktur erhalten geblieben. Da sich die tertiäre Struktur der sieben Sekundärtransporter so stark ähnelt, ist zu vermuten, dass auch der Transportmechanismus ähnlich, jedoch nicht identisch ist. Nach dem strukturellen Aufbau der Transporter, der Lage der Substratbindestellen in den jeweiligen Transportern und der Tatsache, dass es sich bei diesen Proteinen um Membranproteine handelt, wurde ein Transportmechanismus aufgestellt, in dem die Bindestelle des zu transportierende Substrats alternierend zu beiden Seiten der Membran zugänglich ist, ohne jedoch jemals den Substratweg innerhalb des Proteins vollständig zu öffnen. Dieser Mechanismus wurde als "alternating access mechanism" beschrieben. Anhand der unterschiedlichen Zustände, in denen einige der Transporter kristallisierten, kann abgeleitet werden, welche Konformationsänderungen erforderlich sind um das Substrat von einer Seiter der Membran auf die andere zu transportieren. Bisher kristallisierten einzelne der sechs Transporter in der nach außen gerichteten offenen Form, der nach außen gerichteten Form, in der die Substratbindestelle jedoch nicht mehr zugänglich ist, in einer Form, die keine Öffnungspräferenz der Substratbindestelle zu einer Seite der Membran hat und in der nach innen gerichteten Form, in der die Substratbindestelle jedoch nicht geöffnet ist. CaiT kristallisierte in der noch fehlenden Konformation, der nach innen gerichteten Form, in der die Substratbindestelle zugänglich ist. Mit dieser noch fehlenend Konformation kann der Transportzyklus des "alternating access mechanism" vollständig beschrieben werden.

Alle drei CaiT-Homologe kristallisierten in der nach innen gerichteten, offenen Konformation. Im Gegensatz zur EcCaiT-Struktur kristallisierte PmCaiT in der substratungebundenen Form. In der StCaiT-Struktur konnte aufgrund der niedrigen Auflösung kein Substrat nachgewiesen werden. In der EcCaiT-Struktur sind zwei γ-Butyrobetain-Moleküle gebunden. Das erste Molekül wurde in der zentralen Substratbindestelle, der sogenannten Tryptophan-Box bestehend aus vier

Tryptophanen, im Zentrum des Protein lokalisiert. Das zweite γ-Butyrobetain-Molekül wurde in einer Vertiefung an der extrazellulären Proteinoberfläche gefunden. Beide Substrate werden hauptsächlich über Kation-π-Interaktionen zwischen der positiv geladenen quatären Ammoniumgruppe des Substrats und des π-Elektronensystems der Tryptophane in den jeweiligen Bindestellen gebunden.

Eine besondere Eigenschaft von CaiT ist der H^+- bzw. Na^+-unabhängige Substrattransport. Die CaiT-Struktur erklärt warum kein zusätzliches Kation benötigt wird um Substrat zu binden oder zu transportieren. In der EcCaiT-Struktur ist eine wichtige polare nicht-bindende Interaktion zwischen der Carboxylgruppe des γ-Butyrobetains und dem Schwefelatom eines Methionins in der zentrale Bindestelle zu erkennen. Dieses Methionin ist konserviert in den prokaryotischen CaiTs und in den Na^+-unabhängigen eukaryotischen L-Carnitin Transportern (OCTN), jedoch ist es nicht konserviert im Na^+-abhängigen verwandten Glycinbetain Transporter BetP. In BetP ist diese Position des Methionins durch ein Valin ersetzt. Die Mutation des Methionins in CaiT zu Valin ermöglicht zwar immernoch die H^+- bzw. Na^+-unabhängige Bindung des Substrates durch die Tryptophan-Box, jedoch ist der Substrattransport nahezu vollständig zerstört. Eine derart wichtige Substrat-koordinierende Funktion des Schwefelatoms eines Methionins wurde bisher nicht beschrieben.

Eine weitere Stelle, die in H^+- bzw. Na^+-abhängigen Transporter mit H^+ bzw. Na^+ besetzt ist, ist in CaiT von einem positiv geladenen Arginin eingenommen. Eine positive Ladung an dieser Stelle stabilisiert den Bereich im Protein in der Nähe der zentralen Substratbindestelle. Die Mutation des Arginins zu Glutamat in CaiT erzielt eine vollständige Inaktivierung des Substrattransports. Durch Zugabe von Na^+ im Transportansatz kann die Substrattransportaktivität der Glutamat-Mutante jedoch teilweise zurückerlangt werden.

Diese eben beschriebenen Aminosäurereste in den beiden Stellen des Proteins erklären die Kationenunabhängigkeit von CaiT.

Die Aktivierung des Antiportmechanismus in CaiT wurde mit Hilfe von Bindungsstudien an rekonstituiertem Protein ermittelt. Diese Messungen ergaben für das Wildtypprotein ein sigmoidales Substratbindungsverhalten, was auf ein positivkooperatives Bindungsverhalten hindeutet. Die beiden Substratbindestellen im Protein sowie die beiden unterschiedlichen Substrate, L-Carnitin und γ-Butyrobetain, lassen auf einen heterotropen positiv-kooperativen Bindungs- und einen allosterisch regulierten Transportmechanismus schließen. Bei diesem Mechanismus erhöht die Bindung eines Substrats in der regulatorischen Bindestelle durch induzierte Konformationsänderungen die Affinität eines anderen Substrats in einer weiteren Substratbindestelle. Die regulatorische Bindestelle in CaiT befindet sich an der extrazellulären Proteinoberfläche. Eine Schwächung der Substrataffinität in dieser Bindestelle durch Einführung einer Mutation, verstärkt das sigmoidale Substratbindungsverhalten und hat einen negativen Einfluss auf den Substrattransport.

Durch die in dieser Arbeit gelösten 3D-Röntgenkristallstrukturen der zwei CaiT-Homologen, PmCaiT und EcCaiT, sowie den durchgeführten funktionellen Studien sowohl an Wildtypprotein wie auch an Mutanten konnte ein L-Carnitin/γ-Butyrobetain Antiport-Mechanismus für CaiT vorzuschlagen werden.

Table of Contents

ZUSAMMENFASSUNG I

TABLE OF CONTENTS VII

LIST OF SYMBOLS AND ABBREVIATIONS XII

LIST OF FIGURES XVI

LIST OF TABLES XIX

1 INTRODUCTION 1
   1.1 CARNITINE 1
      1.1.1 *Carnitine metabolism in eukaryotes* 1
      1.1.2 *L-carnitine used as osmo- and cryoprotectant* 2
      1.1.3 *Carnitine metabolism in microorganisms* 3
         1.1.3.1 Degradation of carnitine 3
         1.1.3.2 Reduction of carnitine to γ-butyrobetaine 4
   1.2 THE PROKARYOTIC CARNITINE/Γ-BUTYROBETAINE ANTIPORTER CAIT 8
   1.3 THE BETAINE/CARNITINE/CHOLINE TRANSPORTER (BCCT) FAMILY 12
   1.4 3D X-RAY STRUCTURES 18
      1.4.1 *Atomic resolution structure of BetP and structurally homologous secondary transporters* 18
         1.4.1.1 Cation binding in LeuT-type transporters 22
   1.5 AIMS OF THIS WORK 25

2 MATERIALS AND METHODS 27
   2.1 MATERIALS 27
      2.1.1 *Instruments* 27
      2.1.2 *Chemicals* 27
      2.1.3 *Reagent kits* 27
      2.1.4 *Column materials* 28
      2.1.5 *Media and antibiotics* 28
         2.1.5.1 LB-Medium 28
         2.1.5.2 2 × YT Medium (with 5 M NaOH to pH 7 adjusted) 28
         2.1.5.3 TB-Medium 28
         2.1.5.4 *SelenoMet* Medium (Molecular Dimensions) 29

	2.1.5.5	Antibiotics	29
2.1.6		*E. coli strains*	*29*
2.1.7		*Oligonucleotide primers*	*30*
2.1.8		*Mutants*	*33*
2.1.9		*Crystallization screens*	*34*
2.2		MOLECULAR BIOLOGICAL METHODS	35
2.2.1		*Polymerase chain reaction*	*35*
2.2.2		*Site-directed mutagenesis*	*36*
2.2.3		*DNA cleavage using restriction endonucleases*	*37*
2.2.4		*Agarose gel electrophoresis*	*37*
2.2.5		*DNA concentration and purity*	*38*
2.2.6		*Ligation of DNA fragments*	*38*
2.2.7		*Preparation and transformation of chemically competent cells*	*39*
2.2.8		*Isolation of vector DNA*	*40*
2.2.9		*Preparation of bacteria glycerol stocks*	*40*
2.3		BIOCHEMICAL METHODS	41
2.3.1		*Protein production of native CaiT*	*41*
2.3.2		*Protein production of selenomethionine-labeled protein*	*42*
	2.3.2.1	Flask based production of SeMet-labeled protein	42
	2.3.2.2	Fed-batch fermentation	43
2.3.3		*Cell disruption*	*44*
	2.3.3.1	Enzymatic cell disruption	44
	2.3.3.2	Cell disruption using a microflidizer	45
2.3.4		*Protein purification*	*45*
	2.3.4.1	Membrane preparation	45
	2.3.4.2	Protein solubilization	46
	2.3.4.3	Immobilized Metal Ion Affinity Chromatography (IMAC)	47
	2.3.4.4	Size exclusion chromatography (SEC)	48
	2.3.4.5	Proteolytic cleavage of the fusion tag	49
	2.3.4.6	Protein concentration	49
2.3.5		*Detergent concentration*	*50*
2.3.6		*Protein concentration*	*50*
	2.3.6.1	Bradford assay	50
	2.3.6.2	Protein denaturing method	51
	2.3.6.3	Amido Black method	51
2.3.7		*Polyacrylamid gel electrophoresis (PAGE)*	*53*
	2.3.7.1	Blue-Native (BN)-PAGE	53
	2.3.7.2	Denaturing SDS-PAGE	54

	2.3.8	Western blot analysis	55
	2.3.9	Thin-layer chromatography (TLC)	57
	2.3.10	Protein reconstitution into liposomes	57
	2.3.11	Transport measurements	58
2.4	**BIOPHYSICAL METHODS**		59
	2.4.1	Fluorescence	59
	2.4.1.1	Substrate binding assays	60
	2.4.2	Freeze-fracture electron microscopy	60
	2.4.3	X-ray crystallography	61
	2.4.3.1	Prediction of 3D crystallization feasibility	61
	2.4.3.2	Crystallization of CaiT	62
	2.4.3.3	Seeding	65
	2.4.3.4	Detergent screens	65
	2.4.3.5	Additive screens	66
	2.4.3.6	Heavy atom screens	66
	2.4.3.7	Cryocrystallography	67
	2.4.4	Principles of X-ray crystallography	68
	2.4.5	Data collection	70
	2.4.6	Data processing	70
	2.4.7	Phasing	71
	2.4.7.1	The phase problem	71
	2.4.7.2	Methods to solve the phase problem	72
	2.4.8	Phase improvement, model building and refinement	77
	2.4.9	Figures	79

3 RESULTS 81

3.1	**EXPRESSION AND PURIFICATION**		82
	3.1.1	Expression of CaiT in E. coli	82
	3.1.1.1	Native CaiT protein	82
	3.1.1.2	SeMet-labelled CaiT protein	83
	3.1.2	Purification of CaiT	84
	3.1.2.1	Native CaiT protein	84
	3.1.2.2	SeMet-labelled CaiT	86
	3.1.3	Sample quality of CaiT	88
	3.1.3.1	Blue-Native PAGE	88
	3.1.3.2	Analytical size exclusion chromatography (SEC)	90
3.2	**3D CRYSTALLIZATION**		92
	3.2.1	3D crystallization feasibility prediction of CaiT	92
	3.2.2	3D crystallization trials	97

3.3	Structure determination	104
3.3.1	Data collection and data processing	104
3.3.2	Phase determination, model building and refinement	108
3.4	The CaiT structure	122
3.4.1	CaiT topology	122
3.4.2	The CaiT trimer	124
3.4.3	Cytoplasmic substrate pathway, central transport site and periplasmic substrate-binding site	126
3.4.3.1	The central transport site of CaiT	128
3.4.3.2	The second substrate binding site of CaiT	130
3.4.3.3	The extracellular substrate pathway to the central binding site	131
3.4.3.4	Replacement of Na^+ ions in the Na^+-independent CaiT	132
3.5	Lipid analysis and reconstitution of CaiT	135
3.5.1	Lipid analysis	135
3.5.2	Reconstitution of CaiT into liposomes	136
3.6	Substrate transport and binding studies	140
3.6.1	Substrate transport studies	140
3.6.1.1	Wildtype CaiT	140
3.6.1.2	PmCaiT mutants	143
3.6.2	Binding studies	147
3.6.2.1	Binding studies of CaiT in detergent solution	147
3.6.2.1.1	Wildtype CaiT	147
3.6.2.1.2	PmCaiT mutants	153
3.6.2.2	Binding studies of reconstituted CaiT	155

4 DISCUSSION 159

4.1	Crystallization behaviour of the three CaiT homologs	159
4.2	Oligomeric state of CaiT	161
4.3	The CaiT structure	163
4.3.1	CaiT trimer architecture	163
4.3.2	The CaiT protomer	165
4.3.2.1	Comparison of a recently published *E. coli* CaiT structure	168
4.3.3	Transporter conformation	171
4.3.3.1	Alternating access models	174
4.4	The CaiT transport mechanism	178
4.4.1	Cation-π interaction	178
4.4.1.1	Cation-π interaction in CaiT	179
4.4.2	Na^+-independent and cooperative substrate/product antiport in CaiT	182

		4.4.2.1	Sodium-independent substrate transport mechanism	182
		4.4.2.2	Cooperative activation of CaiT	187
		4.4.2.3	Substrate translocation model	190
	4.5	CONCLUDING REMARKS AND OUTLOOK		195

5 LITERATURE 197

6 APPENDIX 211

   6.1 PROTEIN CONSTRUCTS 211

   6.2 VECTOR SYSTEM 213

   6.3 SEQUENCE ALIGNMENT 214

   6.4 CRYSTAL PACKING 217

7 ACKNOWLEDGEMENTS 221

CURRICULUM VITAE 224

List of symbols and abbreviations

°C	degrees Celsius
2D	two-dimensional
3D	three-dimensional
Å	Angstrom(s)
AdiC	arginine/agmatine antiporter
Amp	ampicilin
APC	amino acids, polyamines and organic cation transporters
ApcT	proton-coupled amino acid transporter
APS	ammonium persulfate
BCCT	betaine/carnitine/choline transporter
BetP	glycine betaine permease
bp	base pairs
BSA	bovine serum albumin
CaiA	crotonoetainyl-CoA reductase
CaiB	CoA transferase
CaiC	betaine:CoA ligase
CaiD	enoyl-CoA dehydratase
CaiT	carnitine/γ-butyrobetaine antiporter
Cam	chloramphenicol
CL	cardiolipin
cmc	critical micelle concentration
CoA	co-enzyme A
Cymal-5	5-Cyclohexyl-1-pentyl-β-D-maltoside
Da	Dalton
DDM	n-Dodecyl-β-D-maltoside
DLPC	1,2-dilauroyl-sn-glycero-3-phosphochoine
DMPC	1,2-dimyristoyl-sn-glycero-3-phosphocholine
DNA	desoxyribonucleic acid
dNTP	deoxyribonucleotide triphosphate

E. coli	*Escherichia coli*
EcCaiT	*Escherichia coli* CaiT
EDTA	ethylenediaminetetraacetic acid
EPL	*E. coli* polar lipids
ESRF	European Synchrotron Radiation Facility
et al	*et alia*
FixAB	flavoprotein homologs
FixC	Ubiquinone oxidoreductase homolog
FixX	novel type of ferredoxin
g	acceleration of gravity or gram(s)
h	hour(s)
HEPES	2-(4-(2-hydroxyethyl)-1-piperazinyl)-ethansulfonacid
His$_6$	hexa-histidine tag
IPTG	Isopropyl-β-D-thiogalactopyranosid
k	kilo
K	Kelvin degrees
l	litre
LB	Luria Bertani
LeuT	sodium-dependent leucine transporter
LPR	lipid-to-protein ratio
M	moles per litre
m	milli
Mhp1	benzyl-hydantoin transporter
min	minute(s)
MW	molecular weight
NCS	non-crystallographic symmetry
NCS-1	nucleobase cation symporter-1 family
NSS	neurotransmitter sodium transporters
PAGE	Polyacrylamide gel gelectrophoresis
PC	phosphatidylcholine
PCR	polymerase chain reaction
PDB	protein data base

PE	phosphatidylethanolamine
Pfu	*Pyrococcus furiosus*
PG	phosphatidylglycerol
pI	isoelectric point
PmCaiT	*Proteus mirabilis* CaiT
POPC	1-palmitoyl-1,2-oleoyl-*sn*-glycero-3-phosphocholine
rmsd	root mean square deviation
rpm	rotations per minute
RT	room temperature
SDS	sodiumdodecylsulfate
SEC	size exclusion chromatography
sec, s	second(s)
SLS	Swiss light source
SSS	solute sodium symporters
StCaiT	*Salmonella typhimurium* CaiT
T	temperature
t	time
Taq	*Thermus aquaticus*
TCEP	tris(2-carboxyethyl)phosphine
TLC	thin layer chromatography
TM	transmembrane
Tris	2-amino-2-hydroxymethyl-propane-1,3-diol
TSS	transformation and storage solution
v/v	volume per volume
vSGLT	*Vibrio parahaemolyticus* sodium/galactose symporter
w/v	weight per volume
w/w	weight per weight
wt	wildtype
µ	micro

Amino acid	One-letter code	Three-letter code
alanine	A	Ala
cysteine	C	Cys
glutamic acid	E	Glu
phenylalanine	F	Phe
glycine	G	Gly
histidine	H	His
isoleucine	I	Ile
lysine	L	Lys
leucine	L	Leu
methionine	M	Met
asparagine	N	Asn
proline	P	Pro
glutamine	Q	Gln
arginine	R	Arg
serine	S	Ser
threonine	T	Thr
valine	V	Val
tryptophan	W	Trp
tyrosine	Y	Tyr
variable	X	

List of Figures

	PAGE
FIGURE 1-1 | Degradation of L-carnitine	4
FIGURE 1-2 | Reduction of L-carnitine to Γ-butyrobetaine	7
FIGURE 1-3 | Model of the CaiT exchange mechanism	9
FIGURE 1-4 | Secondary structure prediction of the *E. coli* CaiT	10
FIGURE 1-5 | 2D averages of CaiT single particles	11
FIGURE 1-6 | 2D crystal lattic, power spectrum and projection map of negatively stained 2D CaiT crystals	11
FIGURE 1-7 | Side view of BetP monomer A	14
FIGURE 1-8 | Sequence alignment of selected BCCT family members	17
FIGURE 1-9 | BetP topology	18
FIGURE 1-10 | Conformations of LeuT-type transporters	20
FIGURE 1-11 | Substrate-binding site of BetP	21
FIGURE 1-12 | Cation positions in LeuT-type transporters	22
FIGURE 1-13 | Proposed sodium binding in BetP	23
FIGURE 2-1 | Fed-batch fermentation diagram	44
FIGURE 2-2 | Physical basis of fluorescence	59
FIGURE 2-3 | Phase diagram for protein crystal growth	63
FIGURE 2-4 | Schematic representation of the hanging drop and sitting drop vapour diffusion method	64
FIGURE 2-5 | Conditions for constructive interference	69
FIGURE 2-6 | Vector diagram for a Friedel pair	73
FIGURE 2-7 | Vector diagram for anomalous scattering conditions	74
FIGURE 2-8 | Fluorescence scan of a Lu-containing CaiT crystal	75
FIGURE 2-9 | Schematic description of the MR method	77
FIGURE 3-1 | SDS-PAGE gel and Western blot of PmCaiT test expression	82
FIGURE 3-2 | SDS-PAGE gel of a PmCaiT purification	85
FIGURE 3-3 | SDS-PAGE gel of purified EcCaiT and StCaiT	86
FIGURE 3-4 | SDS-PAGE gel of purified SeMet-PmCaiT and SeMet-EcCaiT	87
FIGURE 3-5 | BN-PAGE gradient gel of purified native and SeMet-labelled CaiT	89
FIGURE 3-6 | Analytical size exclusion chromatography of CaiT	91
FIGURE 3-7 | Prediction of EcCaiT crystallization probability	94
FIGURE 3-8 | Prediction of PmCaiT crystallization probability	95
FIGURE 3-9 | Prediction of StCaiT crystallization probability	96
FIGURE 3-10 | Initial crystallization trials of EcCaiT, PmCaiT and StCaiT	98
FIGURE 3-11 | SDS-PAGE gel and BN-PAGE gradient gel of dissolved CaiT crystals	99
FIGURE 3-12 | Optimized PmCaiT crystals	101

FIGURE 3-13 | OPTIMIZED ECCAIT CRYSTALS 102
FIGURE 3-14 | OPTIMIZED STCAIT CRYSTALS 103
FIGURE 3-15 | DIFFRACTION PATTERN OF A PMCAIT CRYSTAL 105
FIGURE 3-16 | DIFFRACTION PATTERN OF A ECCAIT CRYSTAL 106
FIGURE 3-17 | DIFFRACTION PATTERN OF AN STCAIT CRYSTAL 107
FIGURE 3-18 | ELECTRON DENSITY MAPS AND CRYSTAL PACKING OF PMCAIT 110
FIGURE 3-19 | RAMACHANDRAN PLOT AND STATISTICS OF THE REFINED PMCAIT MODEL 112
FIGURE 3-20 | ELECTRON DENSITY MAPS AND CRYSTAL PACKING OF ECCAIT 114
FIGURE 3-21 | RAMACHANDRAN PLOT AND STATISTICS OF THE REFINED ECCAIT MODEL 116
FIGURE 3-22 | ELECTRON DENSITY MAPS AND CRYSTAL PACKING OF STCAIT 118
FIGURE 3-23 | RAMACHANDRAN PLOT AND STATISTICS OF THE REFINED STCAIT MODEL.................... 120
FIGURE 3-24 | CAIT TOPOLOGY OF THE TWO 5-TM INVERTED REPEATS MOTIF 123
FIGURE 3-25 | CAIT HOMOTRIMER VIEWED FROM THE PERIPLASMIC SIDE AND PARALLEL TO THE
    MEMBRANE 124
FIGURE 3-26 | CAIT PROTOMER–PROTOMER INTERACTION AND PROTOMER–DETERGENT INTERACTION
    125
FIGURE 3-27 | UNOBSTRUCTED CYTOPLASMIC SUBSTRATE PATHWAY OF CAIT 126
FIGURE 3-28 | SLICE THROUGH THE ECCAIT PROTOMER, VIEWED PARALLEL TO THE MEMBRANE 127
FIGURE 3-29 | STEREO VIEW OF THE CENTRAL TRANSPORT SITE OF ECCAIT WITH BOUND SUBSTRATE. 128
FIGURE 3-30 | STEREO VIEW OF THE CENTRAL TRANSPORT SITE OF PMCAIT 129
FIGURE 3-31 | STEREO VIEW OF THE EXTERNAL SUBSTRATE-BINDING SITE OF ECCAIT 130
FIGURE 3-32 | STEREO VIEW OF THE EXTERNAL SUBSTRATE-BINDING SITE OF PMCAIT 131
FIGURE 3-33 | STEREO VIEW OF THE RESIDUES AND HYDROGEN BOND NETWORK BLOCKING THE
    SUBSTRATE PATHWAY ON THE PERIPLASMIC SIDE 132
FIGURE 3-34 | STEREO VIEW OF THE HYDROGEN BOND NETWORK AROUND ARG262 134
FIGURE 3-35 | SILICA GEL TLC PLATE OF *E. COLI* LIPIDS, CYMAL-5, PMCAIT, ECCAIT AND STCAIT .. 136
FIGURE 3-36 | FREEZE-FRACTURE IMAGE OF ECCAIT WT RECONSTITUTED INTO EPL LIPOSOMES 137
FIGURE 3-37 | FREEZE-FRACTURE IMAGES OF PMCAIT WT AND PMCAIT MUTANTS RECONSTITUTED INTO
    EPL LIPOSOMES 138
FIGURE 3-38 | FREEZE-FRACTURE IMAGE OF STCAIT WT RECONSTITUTED INTO EPL–LIPOSOMES 139
FIGURE 3-39 | L-CARNITINE UPTAKE INTO PROTEOLIPOSOMES WITH RECONSTITUTED PMCAIT, ECCAIT
    OR STCAIT 141
FIGURE 3-40 | L-CARNITINE UPTAKE INTO PROTEOLIPOSOMES WITH RECONSTITUTED PMCAIT OR
    ECCAIT IN THE PRESENCE OF NA^+ 142
FIGURE 3-41 | SUBSTRATE TRANSPORT ACTIVITY OF PMCAIT MUTANTS 144
FIGURE 3-42 | BN-PAGE GRADIENT GEL OF PMCAIT E111A AND PMCAIT WT 145
FIGURE 3-43 | NA^+-DEPENDENT SUBSTRATE TRANSPORT OF PMCAIT ARG262GLU 146
FIGURE 3-44 | INFLUENCE OF POTENTIAL SUBSTRATES ON THE INTRINSIC FLUORESCENCE OF PMCAIT WT,
    ECCAIT WT AND STCAIT WT 148

XVII

FIGURE 3-45 | Γ-BUTYROBETAINE INDUCED FLUORESCENCE CHANGE OF PMCAIT WT, ECCAIT WT AND STCAIT WT 149
FIGURE 3-46 | L-CARNITINE INDUCED FLUORESCENCE CHANGE OF PMCAIT WT, ECCAIT WT AND STCAIT WT 150
FIGURE 3-47 | BINDING OF Γ-BUTYROBETAINE OR L-CARNITINE TO PMCAIT WT, ECCAIT WT OR STCAIT WT, MONITORED BY TRYPTOPHAN FLUORESCENCE 151
FIGURE 3-48 | Na^+- OR pH-INDEPENDENT L-CARNITINE BINDING OF PMCAIT WT 152
FIGURE 3-49 | SUBSTRATE BINDING ACTIVITY OF PMCAIT M331V AND PMCAIT W316A 154
FIGURE 3-50 | BINDING STUDIES OF RECONSTITUTED PMCAIT WT, PMCAIT TRP316ALA AND ECCAIT WT 157
FIGURE 4-1 | 3D CAIT STRUCTURE MODELED INTO 2D AVERAGES OF SINGLE PARTICLES 162
FIGURE 4-2 | CAIT TRIMER WITH SURFACE POTENTIAL SHOWN FOR ONE OF THE PROTOMERS 164
FIGURE 4-3 | STEREO VIEW OF THE ECCAIT PROTOMER 166
FIGURE 4-4 | PROPOSED MECHANISM OF THE TRANSITION FROM THE OUTWARD-FACING TO THE INWARD-FACING CONFORMATION IN LEUT-TYPE TRANSPORTERS 174
FIGURE 4-5 | CONFORMATIONAL CHANGES OF LEUT PREDICTED BY THE ROCKING BUNDLE MODEL 176
FIGURE 4-6 | CATION-Π INTERACTION 178
FIGURE 4-7 | GEOMETRY OF CATION-Π INTERACTIONS 179
FIGURE 4-8 | GEOMETRY OF THE CATION-Π INTERACTION IN THE CENTRAL TRANSPORT SITE OF ECCAIT 180
FIGURE 4-9 | GEOMETRY OF THE CATION-Π INTERACTION IN THE REGULATORY BINDING SITE OF ECCAIT 180
FIGURE 4-10 | PUTATIVE CENTRAL TRANSPORT SITE IN APCT 184
FIGURE 4-11 | POSITION OF NA2 IN LEUT AND THE EQUIVALENT POSITION IN THE Na^+-INDEPENDENT TRANSPORTERS CAIT AND APCT 186
FIGURE 4-12 | SUPERPOSITION OF THE PMCAIT TRIMER ON THE ECCAIT TRIMER 189
FIGURE 4-13 | PROPOSED SUBSTRATE ANTIPORT MECHANISM 191
FIGURE 4-14 | IRIS-LIKE HELIX MOVEMENT IN THE TRANSITION FROM THE INWARD-FACING, OPEN CONFORMATION TO THE OCCLUDED CONFORMATION 192
FIGURE 4-15 | CONFORMATIONAL CHANGES IN THE TRANSITION FROM THE INWARD-FACING, OPEN TO THE OUTWARD-FACING, OPEN CONFORMATION IN CAIT 194
FIGURE 4-16 | ECCAIT PROTOMERS 168
FIGURE 4-17 | SUBSTRATE MOLECULES IN THE SUBSTRATE PATHWAY OF ECCAIT 169
FIGURE 4-18 | CENTRAL TRANSPORT SITE AND REGULATORY BINDING SITE 170
FIGURE 6-1 | PET-15B VECTOR MAP 213
FIGURE 6-2 | SEQUENCE ALINGMENT 216
FIGURE 6-3 | UNIT CELL PACKING OF PMCAIT 217
FIGURE 6-4 | UNIT CELL PACKING OF ECCAIT 217
FIGURE 6-5 | UNIT CELL PACKING OF STCAIT 218

List of Tables

PAGE

TABLE 1-1 | BCCT FAMILY MEMBERS 13
TABLE 2-1 | PRIMER FOR THE PM*CAIT* CONSTRUCT 30
TABLE 2-2 | PRIMER FOR THE EC*CAIT* CONSTRUCT 32
TABLE 2-3 | PRIMER FOR THE ST*CAIT* CONSTRUCT 33
TABLE 2-4 | MUTANTS OF PMCAIT 33
TABLE 2-5 | PCR MIXTURE 35
TABLE 2-6 | STANDARD PCR TEMPERATURE PROGRAM 36
TABLE 2-7 | TEA BUFFER 37
TABLE 2-8 | DNA SAMPLE BUFFER 38
TABLE 2-9 | TSS MEDIUM 39
TABLE 2-10 | MEMBRANE BUFFER 46
TABLE 2-11 | SOLUBILIZATION BUFFER 46
TABLE 2-12 | BUFFERS USED FOR AFFINITY CHROMATOGRAPHY PURIFICATION 48
TABLE 2-13 | BUFFERS USED FOR PREPARATIVE AND ANALYTIC SEC 49
TABLE 2-14 | SOLUTIONS FOR THE AMIDO BLACK ASSAY 52
TABLE 2-15 | BN-PAGE BUFFERS 53
TABLE 2-16 | SDS-PAGE GEL MIXTURE FOR TWO GELS 54
TABLE 2-17 | SDS-PAGE RUNNING BUFFER 55
TABLE 2-18 | 4× SAMPLE BUFFER 55
TABLE 2-19 | COOMASSIE STAINING SOLUTIONS 55
TABLE 2-20 | BUFFERS FOR WESTERN BLOT ANALYSIS 56
TABLE 2-21 | OPTIMIZED CRYSTALLIZATION CONDITIONS 65
TABLE 3-1 | DATA COLLECTION STATISTICS 108
TABLE 3-2 | KINETIC ANALYSIS OF PMCAIT WT, ECCAIT WT AND STCAIT WT 143
TABLE 3-3 | SUBSTRATE TRANSPORT ACTIVITY CONSTANTS OF PMCAIT MUTANTS 146
TABLE 3-4 | SUBSTRATE BINDING CONSTANTS FOR PMCAIT WT, ECCAIT WT AND STCAIT WT 153
TABLE 3-5 | SUBSTRATE BINDING CONSTANTS OF PMCAIT M331V AND PMCAIT W316A 155
TABLE 3-6 | SUBSTRATE BINDING CONSTANTS OF PMCAIT WT, PMCAIT W316A AND ECCAIT WT
    RECONSTITUTED INTO LIPOSOMES 158
TABLE 4-1 | STRUCTURE COMPARISONS OF PMCAIT AND ECCAIT TO EACH OTHER AND TO THE
    STRUCTURALLY RELATED TRANSPORTERS BETP, VSGLT, LEUT AND MHP1 172
TABLE 4-2 | LOOKUP TABLE OF RESIDUES FOR THE 10 TM HELICES OF THE TWO INVERTED REPEATS IN
    LEUT-TYPE TRANSPORTERS 173

1 Introduction

1.1 Carnitine

L-carnitine (R-(-)-3-hydroxy-4-trimethylaminobutyrate) is biosynthesized from the amido acids lysine and methionine (Horne and Broquist, 1973; Kleber, 1997; Tanphaichitr and Broquist, 1973). It is an essential metabolite in animals, plants, and prokaryotes. L-carnitine is used both as carbon and nitrogen source in prokaryotes grown under aerobic, as well as in some bacteria grown under anaerobic conditions. In some bacteria it is required as electron acceptor, osmoprotectant or cryoprotectant. In eukaryotic cells, L-carnitine is needed as a carrier of acyl groups. In mammalian cells, the main use of L-carnitine is in mitochondrial β-oxidation of long-chain fatty acids.

1.1.1 Carnitine metabolism in eukaryotes

Carnitine homeostasis in mammals is maintained by endogenous synthesis, absorption from dietary sources and highly efficient tubular reabsorption in the kidney (Rebouche and Seim, 1998). Carnitine is an essential metabolite with numerous roles in the metabolism of liver, skeleton muscle and heart, kidney, and brain (Bremer, 1983). Carnitine has a crucial function in the mitochondrial shuttle system for activated long-chain fatty acids. Fatty acyl carnitine diffuses from the mitochondrial intermembrane space across the inner membrane into the mitochondrial matrix, where β-oxidation takes place (Ramsay et al., 2001; Vaz and Wanders, 2002).

Although animals obtain carnitine primarily from the diet, additional endogenous synthesis of carnitine contributes to carnitine homeostasis. Carnitine is synthesized in mammals from amino acids. The carbon backbone of carnitine is

Introduction

provided by lysine while the 4-N-methyl groups originate from methionine (Rebouche and Seim, 1998; Vaz and Wanders, 2002).

In humans, approximately 70 – 80 % of carnitine is absorbed from the food supply. The carnitine concentration in tissues such as heart, muscle, liver and kidney is generally 20 to 50-fold higher than in the plasma (Tein, 2003). Most of the tissues depend on carnitine uptake from the blood by active transport (Vaz and Wanders, 2002). A subfamily of the organic cation transporter family, the carnitine/cation OCTN transporters OCTN1, OCTN2 and OCTN3 have been isolated and characterized in mice (Tamai *et al.*, 2000). The three carntitine transporters of the OCT subfamily exhibit a sequence identity of 75 %. Carnitine transport by OCTN1 was shown to be H^+- dependent (Tamai *et al.*, 1997), OCTN2 showed Na^+-dependence (Tamai *et al.*, 2001) whereas the carnitine transport by OCTN3 was shown to be H^+- and Na^+-independent (Durán *et al.*, 2005). However, the electrochemical gradient dependence of carnitine/cation transporters described in the literature is inconsistent, which may be attributable to interspecies variation, tissue-specific kinetic differences and differences in tissue preparation or methodology (Tein, 2003).

Studies support that mutations in the OCTN gens (SLC22A4 encoding OCTN1 and SLC22A5 encoding OCTN2) confer risk for Crohn's disease (Noble *et al.*, 2005; Peltekova *et al.*, 2004). Crohn's disease is a chronic, inflammatory disorder of the gastrointestinal tract. SLC22A4 and SLC22A5 are widely expressed (Elimrani *et al.*, 2003; Wu *et al.*, 2000; Wu *et al.*, 1999) but they are specifically expressed in intestinal cell types that are infected by Crohn's disease (Peltekova *et al.*, 2004). Defects in the organic carnitine/cation transporters may diminish uptake of physiologic compounds while increasing uptake of potential toxins and may impair fatty acid β-oxidation in intestinal epithelium (Peltekova *et al.*, 2004).

1.1.2 L-carnitine used as osmo- and cryoprotectant

L-carnitine is involved in osmoprotection in *E. coli* (Jung *et al.*, 1990a; Verheul *et al.*, 1998) and other microorganisms (Kleber, 1997), both under aerobic and anaerobic growth conditions. When *E. coli* is exposed to increased osmolarity, it responds by accumulating osmoregulatory solutes to increase the osmotic pressure inside the cell to avoid dehydration (Le Rudulier *et al.*, 1984). These solutes include ions like K^+, amino acids such as proline or glutamate and other zwitterionic organic compounds like glycine betaine (one of the most widely used osmoprotectants), γ-butyrobetaine, crotonobetaine or carnitine (Kappes *et al.*, 1996). The molecular basis for their action as osmoprotectants arises from the fact that they can be accumulated to high concentrations inside the cell in response to increased external osmolarity and thus are able to restore cellular hydration and volume without disrupting cellular functions (Yancey *et al.*, 1982). In *E. coli* the major osmoregulated transporters are ProU, ProP, and BetT. ProU is a glycine betaine and proline betaine ABC transporter (Haardt and Bremer, 1996), which shares sequence homology with the carnitine transporter OpuC from *Bacillus subtilis* and *Listeria monocytogenes*. The H^+/proline betaine symporter ProP (MacMillan *et al.*, 1999) belongs to the major facilitator superfamily (MFS). BetT is a proton-dependent choline transporter (Lamark *et al.*, 1991) belonging to the betaine/carnitine/choline transporter (BCCT) family.

Some bacteria such as *L. monocytogenes* use osmoprotectants as cryoprotectants (Ko *et al.*, 1994). It has been shown that in *L. monocytogenes* carnitine is accumulated very efficiently under both osmotic and chill stress (Smith, 1996). The major carnitine transporter in *L. monocytogenes* is the ABC transporter OpuC (Angelidis *et al.*, 2002).

1.1.3 Carnitine metabolism in microorganisms

1.1.3.1 Degradation of carnitine

Some bacteria (e.g. *Pseudomonas*) are able to use L-carnitine as the sole source of carbon and nitrogen (Kleber, 1997). When these organisms grow under aerobic conditions, supplementation of L-carnitine induces the expression of the enzyme carnitine dehydrogenase (L-carnitine : NAD^+ oxidoreductase), which reduces

Introduction

carnitine to 3-dehydrocarnitine while oxidizing one molecule NAD$^+$ to NADH (Figure 1-1; (Kleber, 1997; Rebouche and Seim, 1998)). 3-Dehydrocarnitine subsequently is degraded to glycine betaine, which is metabolized by step demethylation to glycine (Eichler *et al.*, 1994; Kleber, 1997).

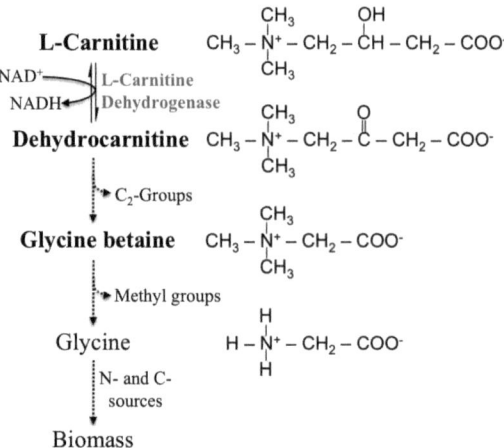

Figure 1-1 | Degradation of L-carnitine
Under aerobic conditions some microorganism use L-carnitine as a sole carbon or nitrogen source. The oxidation of L-carnitine to 3-dehydrocarnitine is catalyzed by the enzyme L-carnitine dehydrogenase. 3-dehydrocarnitine is subsequently cleaved to glycine betaine, which is further degraded to glycine and other biomass precursors (Kleber, 1997; Rebouche and Seim, 1998).

1.1.3.2 Reduction of carnitine to γ-butyrobetaine

Enterobacteria such as *Escherichia coli*, *Proteus mirabilis*, and *Salmonella typhimurium*, do not assimilate the carbon and nitrogen skeleton of trimethylammonium compounds but are able to metabolize carnitine *via* crotonobetaine to γ-butyrobetaine (Figure 1-2; (Kleber, 1997). The carnitine metabolic pathway involves the products of the *caiTABCDE* operon. Transcription of *caiTABCDE* is induced under anaerobic conditions by carnitine or crotonobetaine (Jung *et al.*, 1990b; Jung *et al.*, 1987; Kleber, 1997). The presence of γ-butyrobetaine

Introduction

and electron acceptors like nitrate and fumarate (Kleber, 1997; Obon *et al.*, 1999) represses *caiTABCDE* transcription (Walt and Kahn, 2002). Expression of the *cai* operon produces six proteins. CaiT has been characterized as a carnitine/γ-butyrobetaine exchanger (Jung *et al.*, 2002). CaiA is a crotonobetainyl-CoA reductase, CaiB is a CoA-transferase, CaiC is a betaine-CoA ligase and CaiD is an enoyl-CoA hydratase. The function of CaiE is still unknown although over-production of CaiE was shown to stimulate the carnitine racemase activity and to increase the level of carnitine dehydratase, suggesting a possible role as a cofactor for carnitine pathway enzymes (Eichler *et al.*, 1994; Rebouche and Seim, 1998). A seventh open reading frame in the *cai* operon has been identified (Engemann *et al.*, 2005) at the 3' region adjacent to the *caiTABCDE* operon. The *caiF* gene encodes for a transcription factor (CaiF) that is transcribed in the opposite direction to the *caiTABCDE* operon. In the presence of L-carnitine, CaiF is activated and induces the expression of the *cai* genes.

At the 5' region adjacent to the *E. coli cai* operon, another operon has been identified with four additional open reading frames, the genes *fixABCX* (Eichler *et al.*, 1995). The two promoters of the *cai* and *fix* operons are arranged back-to-back, separated by a 289-bp region which contains binding sites for regulatory proteins (Buchet *et al.*, 1998).

The proteins FixA and FixB of *E. coli* resemble the β- and α-subunit of the electron transfer flavoproteins (Tsai and Saier, 1995) that transfer electrons from a dehydrogenase to ubiquinone oxidoreductase. FixC has regions that resembles the ubiquinone oxidoreductase family (Walt and Kahn, 2002). FixX is predicted to be a novel type of ferredoxin (Bruschi and Guerlesquin, 1988). It has been hypothesized that the FixA and FixB proteins accept electrons from acyl-CoA dehydrogenases and pass them to FixC, which donates electrons *via* FixX to acyl-CoA reductases (Walt and Kahn, 2002).

Under anaerobic growth conditions and in the presence of carbon and nitrogen sources, enterobacteria metabolize carnitine to γ-butyrobetaine in a multi-step process if no other electron acceptors are present (Figure 1-2; (Bernal *et al.*, 2008; Kleber, 1997; Rebouche and Seim, 1998)). The degradation of L-carnitine starts with the transfer of CoA to carnitine. This step is catalyzed by the enzyme betaine:CoA ligase

Introduction

(CaiC), which has been demonstrated to be ATP dependent (Bernal *et al.*, 2008). The activated L-carnitinyl-CoA is further dehydrated by the enzyme enoyl-CoA dehydratase (CaiD; (Jung *et al.*, 1989)) to crotonobetainyl-CoA. The dehydration reaction of L-carnitinyl-CoA to crotonobetainyl-CoA is reversible and the production of L-carnitine from crotonobetaine is possible. The subsequent reduction reaction of crotonobetainyl-CoA to γ-butyrobetainyl-CoA catalyzed by crotonobetainyl-CoA reductase (CaiA) is an irreversible reaction. This reaction step involves the proteins FixA, FixB, FixC and FixX. In the last reaction step where the final product γ-butyrobetaine is produced (Jung *et al.*, 1990b; Kleber, 1997; Seim and Kleber, 1988), CoA is regenerated. The CoA-transferase CaiB catalyzes this last reaction. Under physiological conditions CaiT (Section 1.2) enables the exchange of L-carnitine against γ-butyrobetaine.

Some strains of the genera *Escherichia* and *Proteus* are able to reverse the L-carnitine to γ-butyrobetaine pathway, and are used for commercial production of L-carnitine. During the L-carnitine biotransformation process, waste products of the chemical industry such as D-carnitine or the achiral substrate crotonobetaine are used as inducer in the growth media. Although the L-carnitine metabolism is related to anaerobic growth conditions, the production of L-carnitine is favoured in the presence of oxygen which inhibits the irreversible crotonobetaine reductase reaction (Obon *et al.*, 1999).

Introduction

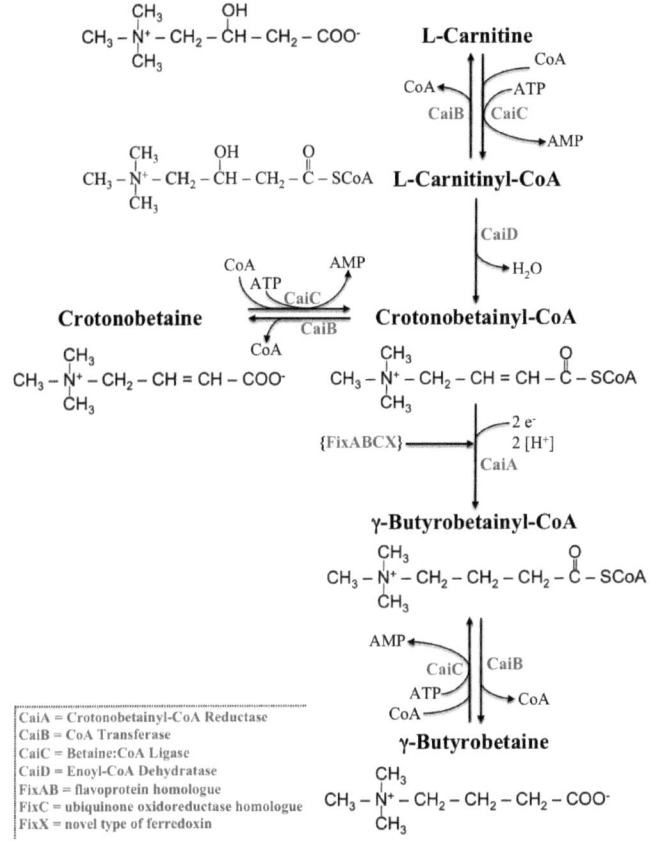

Figure 1-2 | Reduction of L-carnitine to γ-butyrobetaine

Under anaerobic conditions enterobacteria metabolize L-carnitine *via* crotonobetaine to γ-butyrobetaine. The biotransformation occurs at the level of CoA derivatives. L-carnitine is activated by the addition of CoA which is catalyzed by the betaine:CoA ligase (CaiC). L-carnitine-CoA is then dehydrated to crotonobetainyl-CoA, catalyzed by the enzyme enoyl-CoA-dehydratase (CaiD), which is a reversible reaction. The next step is the non-reversible reduction of crotonobetainyl-CoA to γ-butyrobetainyl-CoA by the enzyme crotonobetainyl-CoA reductase (CaiA). The FixABCX system supplies the electrons to CaiA, which are required in the synthesis of γ-butyrobetaine. After CoA activation, crotonobetaine, supplied in the external media, can directly be used as an electron acceptor in the biotransformation of γ-butyrobetaine. The last reaction step is catalyzed by the enzyme CoA transferase (CaiB), which regenerates CoA from γ-butyrobetaine-CoA and γ-butyrobetaine is produced. (Bernal *et al.*, 2008; Rebouche and Seim, 1998; Walt and Kahn, 2002)

1.2 The prokaryotic carnitine/γ-butyrobetaine antiporter CaiT

In enterobacteria such as *E. coli*, *P. mirabilis* and *S. typhimurium*, carnitine is taken up by the secondary transporter CaiT that belongs to the glycine betaine/carnitine/choline transporter (BCCT) family. CaiT is the main transporter of betaine derivatives when, under anaerobic conditions, the transcription of the *caiTABCDE* operon is induced (Canovas *et al.*, 2003; Jung *et al.*, 1990b; Jung *et al.*, 1987; Kleber, 1997). It has been shown that CaiT functions as a specific antiporter (Jung *et al.*, 2002). CaiT exchanges D-/L-carnitine, crotonobetaine and γ-butyrobetaine in both directions across the membrane (Figure 1-3). Unlike its relatives, the Na^+/glycine betaine symporter BetP from *C. glutamicum* (Farwick *et al.*, 1995) and the proton-dependent choline transporter BetT from *E. coli* (Lamark *et al.*, 1991), which are activated by osmotic stress, CaiT is constitutively active and, in contrast to many other secondary transporters, no proton or sodium dependence of substrate binding or transport has been found (Jung *et al.*, 2002).

Introduction

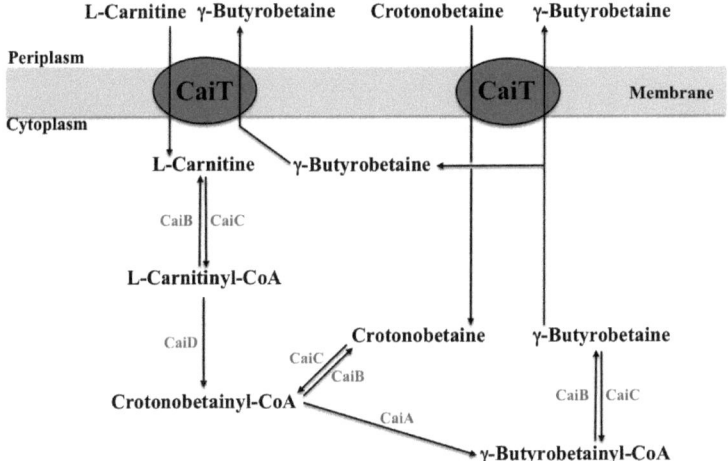

Figure 1-3 | Model of the CaiT exchange mechanism
CaiT is a highly specific exchanger for products involved in the reduction of L-carnitine to γ-butyrobetaine (Figure 1-2). Under native conditions CaiT exchanges L-carnitine against γ-butyrobetaine and, when supplied in the medium, crotonobetaine against γ-butyrobetaine.

The *caiT* gene encodes a protein of 504 – 514 amino acids, depending on the species. Hydropathy analysis of the amino acid sequence predicts 12 putative TM helices, with the amino and carboxy termini on the cytoplasmic side of the membrane (Figure 1-4; (Jung *et al.*, 2002)). The highly conserved amino acids of TM 8 (Figure 1-4, Figure 1-8) are predicted to play an important role in substrate binding and translocation across the membrane (Kappes *et al.*, 1996).

Introduction

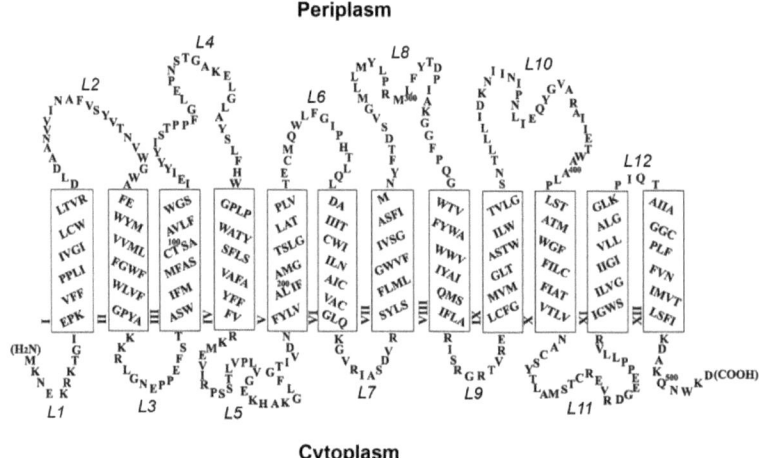

Figure 1-4 | Secondary structure prediction of the *E. coli* CaiT

The carnitine/γ-butyrobetaine antiporter from *E. coli* consists of 504 amino acids. Hydropathy analysis of the CaiT amino acid sequence predicts 12 putative TM helices with the amino and carboxy termini facing the cytoplamic side. The predicted TM helices are represented as rectangles and are numbered I – XII. The predicted loop regions are numbered L1 – L12, starting from the N-terminus (Jung *et al.*, 2002).

Previous structural studies of the L-carnitine/γ-butyrobetaine antiporter from *E. coli* showed CaiT to be a trimer both in detergent solution and in the lipid membrane of two-dimensional crystals (Figure 1-5, Figure 1-6; (Vinothkumar *et al.*, 2006)).

Introduction

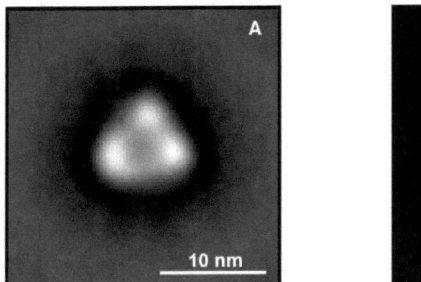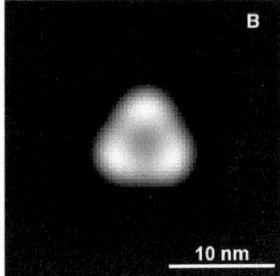

Figure 1-5 | 2D averages of CaiT single particles

The 2D averages of single particles of CaiT solubilized in DDM showing a trimeric structure. No symmetry was applied in (**A**), whereas three-fold symmetry was applied in (**B**) (Vinothkumar *et al.*, 2006).

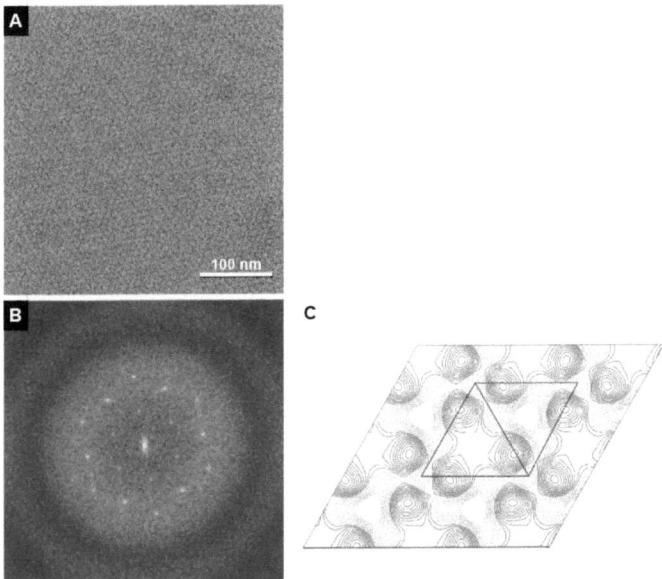

Figure 1-6 | 2D crystal lattic, power spectrum and projection map of negatively stained 2D CaiT crystals

2D crystal lattices of single membranes (**A**) and the corresponding power spectrum (**B**) of negatively stained 2D CaiT crystals. The projection map (**C**) of single membrane 2D crystals was calculated to a resolution of 25 Å with *P3* symmetry applied (Vinothkumar *et al.*, 2006).

Introduction

1.3 The betaine/carnitine/choline transporter (BCCT) family

The BCCT family comprises a large group of secondary transporters (Table 1-1), which share the common feature of transporting substrates containing a quaternary ammonium group [$R - N^+(CH_3)_3$].

So far, the best-characterized member of the BCCT family is the Na^+/glycine betaine symporter BetP from *Corynebacterium glutamicum* (Farwick *et al.*, 1995; Morbach and Krämer, 2005; Peter *et al.*, 1996; Ressl *et al.*, 2009; Ziegler *et al.*, 2004). The 3.35 Å X-ray structure of BetP (Ressl *et al.*, 2009) shows a homotrimeric arrangement of the protein. Each protomer has 12 transmembrane (TM) α–helices, with the amino and carboxy termini on the cytoplasmic side (Figure 1-7). In addition, one α–helix (h7) runs almost parallel to the membrane and two small helices, the intracellular helix IH1 and the extracellular helix EH2, connect the helix pairs TM4 to TM5 and TM9 to TM10, respectively (Figure 1-7). The C-terminal domain of the protomer forms a long α–helix that contains two positively charged clusters, termed +1 and +2 (Ressl *et al.*, 2009).

Introduction

Table 1-1 | BCCT family members

Protein	Main substrates	Organism	Transport Mechanism	Function	Reference
CaiT	carnitine/γ-butyrobetaine	E. coli	Antiport	Redoxprotectant transporter	(Jung et al., 1989); (Eichler et al., 1994); (Jung et al., 2002)
BetT	choline	E. coli	Symport	Osmoprotectant transporter	(Lamark et al., 1991)
BetU	glycine betaine	E. coli	Symport	Osmoprotectant transporter	(Ly et al., 2004)
BetP	glycine betaine	C. glutamicum	Symport	Osmoprotectant transporter	(Farwick et al., 1995); (Peter et al., 1996)
EctP	ectoine	C. glutamicum		Osmoprotectant transporter	(Steger et al., 2004)
LcoP	proline	C. glutamicum		Osmoprotectant transporter	(Steger et al., 2004)
OpuD	glycine betaine	B. subtilis	Symport	Osmoprotectant transporter	(Kappes et al., 1996)
BetM	glycine betaine	M. halophilus	Symport	Osmoprotectant transporter	(Vermeulen and Kunte, 2004)
EctM	ectoine	M. halophilus		Osmoprotectant transporter	(Vermeulen and Kunte, 2004)
ButA	glycine betaine	T. halophila	Symport	Osmoprotectant transporter	(Baliarda et al., 2003)
BetS	glycine betaine	S. melitoli	Symport	Osmoprotectant transporter	(Boscari et al., 2002)
CudT	choline	S. xylosus	Symport	Osmoprotectant transporter	(Rosenstein et al., 1999)
BetL	glycine betaine	L. monocytogenes	Symport	Osmoprotectant transporter	(Sleator et al., 1999)
BetT	choline	P. syringae	Symport	Osmoprotectant transporter	(Chen and Beattie, 2008)
BetT	choline	H. influenzae	Symport	Osmoprotectant transporter	(Fan et al., 2003)

Introduction

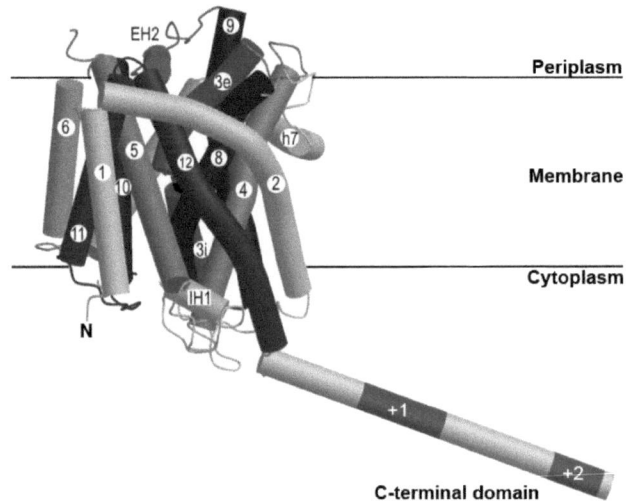

Figure 1-7 | Side view of BetP monomer A
The BetP monomer is built up of 12 TM helices (numbered 1 – 12). The N- and C-termini are located on the cytoplasmic side of the membrane. The long loop regions connecting TM4 with TM5 on the cytoplasmic side and TM9 with TM10 on the periplasmic side form the short helices, which are termed IH1 and EH2. The C-terminal domain forms a long α–helix that contains two positively charged patches (+1 and +2) (Ressl et al., 2009).

Most BCCT members respond to hyperosmotic shock by transporting compatible solutes, which serve as osmoprotectants, into or out of the cell. The most effective and widely used osmoprotectant is glycine betaine. BCCTs involved in glycine betaine transport are listed in Table 1-1. Other BCCTs transport ectoine (EctP in *C. glutamicum* or EctM in *M. halophilus*), proline (LcoP in *C. glutamicum*), carnitine (CaiT in *E. coli*) or choline (BetT in *E. coli*, *P. syringae* and *H. influenzae*) (Table 1-1). In *E. coli*, also choline transport is osmotically activated. However, choline is not directly used as a compatible solute (Styrvold et al., 1986), but is oxidized to the osmoprotectant glycine betaine by the Bet enzymes BetI, BetB and BetA (Tondervik and Strom, 2007).

A sequence alignment of the BCCTs CaiT and BetT from *E. coli*, BetP from *C. glutamicum* and OpuD from *B. subtilis* (Figure 1-8) indicates long C-terminal ends

Introduction

for BetP and BetT. For BetT it was shown that the C-terminal domain is involved in regulation of osmotic activation (Peter *et al.*, 1996; Tondervik and Strom, 2007). The hydrophilic C-terminal domain in BetP (Figure 1-7) is involved in membrane lipid mediated osmosensing and regulation of the BetP transport activity (Krämer and Morbach, 2004; Krämer and Ziegler, 2009). CaiT and OpuD do not have such long C-termini and a regulation by the C-terminus could not be demonstrated. CaiT is exceptional within the BCCT family in that it is the only transporter not involved in the regulation of osmotic stress (Verheul *et al.*, 1998).

With the exception of CaiT, all BCCT family transporters show a Na^+- or H^+-dependent substrate transport that allows high accumulation of osmoprotectant within the cell. In the BetP X-ray structure the substrate glycine betaine was co-crystallized within the binding pocket of the transporter (Ressl *et al.*, 2009). The substrate-binding pocket in BetP is defined by three tryptophans (Trp) and one tyrosine (Tyr). These residues are highly conserved within the BCCT family (Figure 1-8; red boxes). In addition, the positions of two Na^+ ions in BetP were modeled by homology to the structurally related LeuT transporter (Section 1.4.1.1). One of these ions is coordinated by the substrate while the other ion is stabilizing the environment of the substrate-binding site (Ressl *et al.*, 2009).

Introduction

```
                                                                       10
                   ..|....|....|....|....|....|....|....|....|....|....|....|
EcCaiT|    1       --------------------------------------------MKNEKRKTGIEP   12
EcBetT|    1       ---------------------------------------MTDLSHSREKDK----INP   15
CgBetP|    1       MTTSDPNPKPIVEDAQPEQITATEELAGLLENPTNLEGKLADAEEEIILEGEDTQASLNW   60
BsOpuD|    1       --------------------------------------------------MLKHISSVFW   10

                            20         30         40         50         60         70
                   ..|....|....|....|....|....|....|....|....|....|....|....|
EcCaiT|   13       KVFFPPLIIVGIL WLTVRDLDAANVVINAVFSYVTNVHGWAFEWYMVVMLFGWFWLVFG   72
EcBetT|   16       VVFYTSAGLILLFSLTTILFRDFSALMIGRTLDWVSKTFGWYYILLAATLYIVFVV IA S   75
CgBetP|   61       SVIVPALVILVLATVVWGIGFKDSFTNFASSALSAVVDNLGWAFILFGTVFVFFIVVIAAS  120
BsOpuD|   11       IVIA----ITAAAVLWGVISPDSLQNVSQSAQAFITDSFGWYYLLVVSLFVGF LFLIFS   66

                           80         90        100        110        120
                   ..|....|....|....|....|....|....|....|....|....|....|....|
EcCaiT|   73       PYAKKRLG--NEPPEFSTASWIFMMFAS TSAAVLFWG-SIEIYYYISTPPFGLEPNSTG  129
EcBetT|   76       RFGSVKLGPEQSKPEFSLLSWAAMLFAAGIGIDLMFFSVAEPVTQYMQPPEG--AGQTIE  133
CgBetP|  121       KFGTIRLGRIDEAPEFRTVSWISMMFAAGMGIGLMFYGTTEPLTFYRNGVP-GHDEHN--  177
BsOpuD|   67       PIGKIKLGKPDEKPEFGLLSWFAMLFSAGMGIGLVFYGAAEPISHYAISSP-SGETETPQ  125

                          130        140        150        160        170        180
                   ..|....|....|....|....|....|....|....|....|....|....|....|
EcCaiT|  130       AKELGLAYSLFHWGPLPWAIDSFLSVAPAYFFFVRKWEVIRPSSTLVPLVGEKHAKGLFG  189
EcBetT|  134       AARQAMVWTLFHTGLTGWSMYALMGMALGYFSYRYNLPLTIR-SALIPIFGKR-INGPIG  191
CgBetP|  177       -VGVAMSTTMFHMTLHPWAIXAIVGLAIAYSTFRVGRKQLLS-SAFVPLIGEKGAEGWLG  235
BsOpuD|  126       AFRDALRYTFFHMGLHAMAIYAIVAL IAYFGFRKGAPGLIS-STLSPILGDK-VNGPIG  183

                          190        200        210        220        230        240
                   ..|....|....|....|....|....|....|....|....|....|....|....|
EcCaiT|  190       TIVDNFYLVALIFAMGTSLGLATPLVTE MQWLFGIPHTL-QLDAIIIT WIILNAI VA  248
EcBetT|  192       HSVDIAAVIGTIFGIATTLGIGVVQLNYGLSVLFDIDPDSM-AAKAALIALSVIIATISVT 250
CgBetP|  236       KLIDILAIIATVFGTA SLGLGALQIGAGLSAANIIEDPSDWTIVG1V5VLTLAFIFSAI  295
BsOpuD|  184       KAID IAVFATVVGVSTSLGLGATQINGGLNYLFGIPNAFIVQLV-LIIIVTVLFLLSAW  242

                          250        260        270        280        290        300
                   ..|....|....|....|....|....|....|....|....|....|....|....|
EcCaiT|  249        GLQKGVRIASDVRSYLSFLMLGWVFIVSGASFIMNYFTDSVGMLLMYLPRMLFY---TD  305
EcBetT|  251       SGVDKGIRVLSELNVALALGLILFVLFMGDTSFLLNALVLNVGDYVNRFMGMTLN--SFA  308
CgBetP|  296       SGVGKGIQYLSNANMVLAALLAIFVFVVGPTVSILNLLPGSIGNYLSNFFQMAGRTAMSA  355
BsOpuD|  243       SGLGKGIKYLSNTNMVLAGLLMLFMLVVGPTVLIMNSPTDSIGQYIQNIVQMSFR-LTPN  301

                          310        320        330        340        350        360
                   ..|....|....|....|....|....|....|....|....|....|....|....|
EcCaiT|  306       PIAKGGFPQGWMTVFYWAWWVIYAIQMSIFLARISRGRTVREL FGMVLGLTASTWILWTV  365
EcBetT|  309       FDRPVEWMMNNWTLFFWMWWVAWSPFVGLFLARISRGRTIRQFVLGTLIIPFTFTLLWLSV  368
CgBetP|  356       DGTAGEWLGSWTIFFYWMWWISWSPFVGMFLARISRGRTIREFILGVLLVPAGVSTVWFSI  415
BsOpuD|  302       DPEKREWINSWTIFYWMWWISWSPFVGIFIARVSRGRTIREFLIGVLVTP ILTFLWFSI  361

                          370        380        390        400        410        420
                   ..|....|....|....|....|....|....|....|....|....|....|....|
EcCaiT|  366       LGSNTLLLIDKNIINIPN-LIEQYGVARAIIETWAALPLSTATMWGFFIL FIATVTLVN  424
EcBetT|  369       FGNSALYEIIHGGAAFAEEAMVHPER--GFYSLLAQYPAFTFSASVATITGLLFYVTSAD  426
CgBetP|  416       FGGTAIVFE-QNGESIWG----DGAAEEQLFGLLHALPGGQIMGIIAMILLGTFFITSAD  470
BsOpuD|  362       FGVSAMDLQ-QKGAFNVA----KLSTETMLFGTLDHYPLTMVTSILALILIAVFFITSAD  416

                          430        440        450        460        470        480
                   ..|....|....|....|....|....|....|....|....|....|....|....|
EcCaiT|  425       A SYTLAMST REVRDGEEPPLLVRIGWSILVGIGIVLLALGG---LKPIQTAIIAGG   481
EcBetT|  427       SGALVLGNFTSQLKDINSDAPGWLRVFWSVAIGLITLGMLMTN---GISALQNTTVIMGL  483
CgBetP|  471       SASTVMGTMSQHGQ---LEANKWVTAAWGVATAAIGLTLLLSGGDNALSNLQNVTIVAAT  527
BsOpuD|  417       SATFVLGMQTSYGS---LNPANSVKLSWGIIQSAMAAVLLYSGG---LAALQNTAILAAL  470

                          490        500
                   ..|....|....|....|....|....|....|....|....|....|....|....|
EcCaiT|  482       PLFFVNIMVTLSFIKDAKQNWKD-------------------------------------  504
EcBetT|  484       PFSFVIFFVMAGLYKSLKVEDYRRESANRDTAPRPLGLQDRISWKKRLSRLWNYPGITRT  543
CgBetP|  528       PFLFVVIGLMFALVKDLSNDVIYLEYREQQRFNARLARERRVHNEHRKRELAAKRRRERK  587
BsOpuD|  471       PFSIVILLMIASLVQSLS-------------------KERR--EIKKAEKLDKPRSPRV  508

                          550        560        570        580        590        600
                   ..|....|....|....|....|....|....|....|....|....|....|....|
EcCaiT|  504       ------------------------------------------------------------ 504
EcBetT|  544       KQMMETV.YPAMEEVAQELRLRGAYVELKSLPPEEGQQLGHLDLLVHMGEEQNFVYQIWP  603
CgBetP|  588       ASGAGKRR---------------------------------------------------  595
BsOpuD|  509       KKAY-------------------------------------------------------  512
```

Figure 1-8 | Sequence alignment of selected BCCT family members

The amino acid sequence of the carnitine/γ-butyrobetaine antiporter CaiT from *E. coli* (EcCaiT) is aligned with the choline transporter BetT form *E. coli* (EcBetT), the Na^+/ glycine betaine symporter BetP from *C. glutamicum* (CgBetP) and the glycine betaine transporter OpuD from *B. subtilis* (BsOpuD). Residues 604 – 677 of BetT were removed from the sequence alignment.

The red-boxed residues are highly conserved within the BCCT family and have been shown to be involved in substrate binding in BetP.

Introduction

1.4 3D X-ray structures

1.4.1 Atomic resolution structure of BetP and structurally homologous secondary transporters

The 3D X-ray structure of the CaiT relative BetP has been solved recently (Ressl *et al.*, 2009). In the crystal, BetP forms symmetric trimers. Each BetP monomer (Figure 1-7) consists of 12 TM helices. Ten of these form two inverted structural repeats of five TM helices, which build up the transporter core (Figure 1-9).

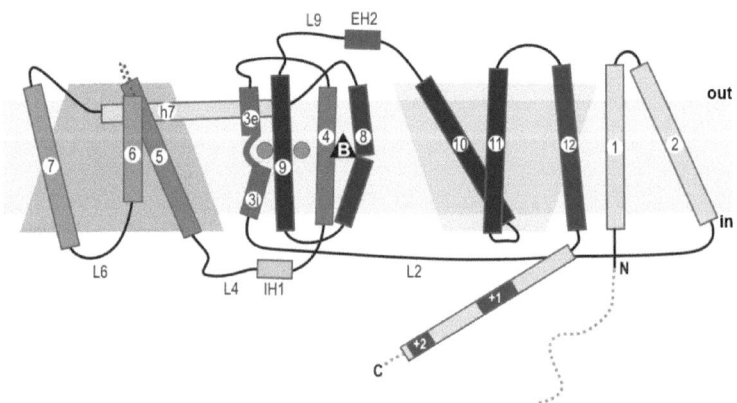

Figure 1-9 | BetP topology
BetP consists of 12 TM helices, with the N- and C-terminus facing the cytoplasmic side of the membrane. The transporter core is formed by two inverted repeats of five TM helices. Repeat 1 (TM4 – TM7) is coloured in orange, with colour intensity decreasing from the N- to the C-terminus. The structurally related repeat 2 (TM8 – TM12) is coloured in blue, again with decreasing colour intensity from the N- to the C-terminus. Protein segments not belonging to the inverted repeats are coloured in grey. The substrate molecule is represented as a black triangle, sodium ions are represented by green spheres (Ressl *et al.*, 2009).

The core architecture of BetP resembles that of several secondary transporters from other families. The first reported crystal structure of a sodium-coupled secondary transporter with two 5-TM inverted repeats was the Na^+/alanine transporter LeuT$_{Aa}$ in complex with leucine from *Aquifex aeolicus* (LeuT-Leu; (Yamashita *et al.*, 2005)), a member of the neurotransmitter/sodium symporter (NSS) family. Since then, five other secondary transporters have revealed the same 5-TM inverted repeat; they are therefore referred as the LeuT-type transporters.

The Na^+/galactose transporter vSGLT from *Vibrio parahaemolyticus* (solute sodium symporter (SSS) family; (Faham *et al.*, 2008)) and the benzyl-hydantoin symporter Mhp1 from *Microbacterium liquefaciens* (nucleobase cation symporter-1 (NCS-1) family; (Weyand *et al.*, 2008)) are sodium-coupled transporters like BetP. The recently solved crystal structures of two proton-coupled transporters of the amino acid, polyamine, organocation (APC) transporter family, AdiC (Fang *et al.*, 2009; Gao *et al.*, 2009; Gao *et al.*, 2010) and ApcT (Shaffer *et al.*, 2009), also show the LeuT-type core architecture.

Comparisons of the various structures revealed four different conformations (Figure 1-10) that provide insights into the substrate transport mechanism with alternating access of the substrate-binding site to either side of the membrane (Jardetzky, 1966). The "outward-facing open" conformation is represented by Mhp1 (Weyand *et al.*, 2008), LeuT-Trp (Singh *et al.*, 2008), and AdiC (Fang *et al.*, 2009; Gao *et al.*, 2009; Gao *et al.*, 2010). LeuT-Leu (Yamashita *et al.*, 2005) and Mhp1 in complex with its substrate benzyl-hydantoin (Mhp1-5FH; (Weyand *et al.*, 2008)) are crystallized in the "outward-facing occluded" conformation, whereas BetP complexed with glycine betaine (BetP-Bet; (Ressl *et al.*, 2009) and ApcT (Shaffer *et al.*, 2009) are trapped in the "occluded" conformation in which the substrate is inaccessible from either side of the membrane. The "inward-facing occluded" conformation is represented by vSGLT with bound galactose (vSGLT-Gal; (Faham *et al.*, 2008).

Introduction

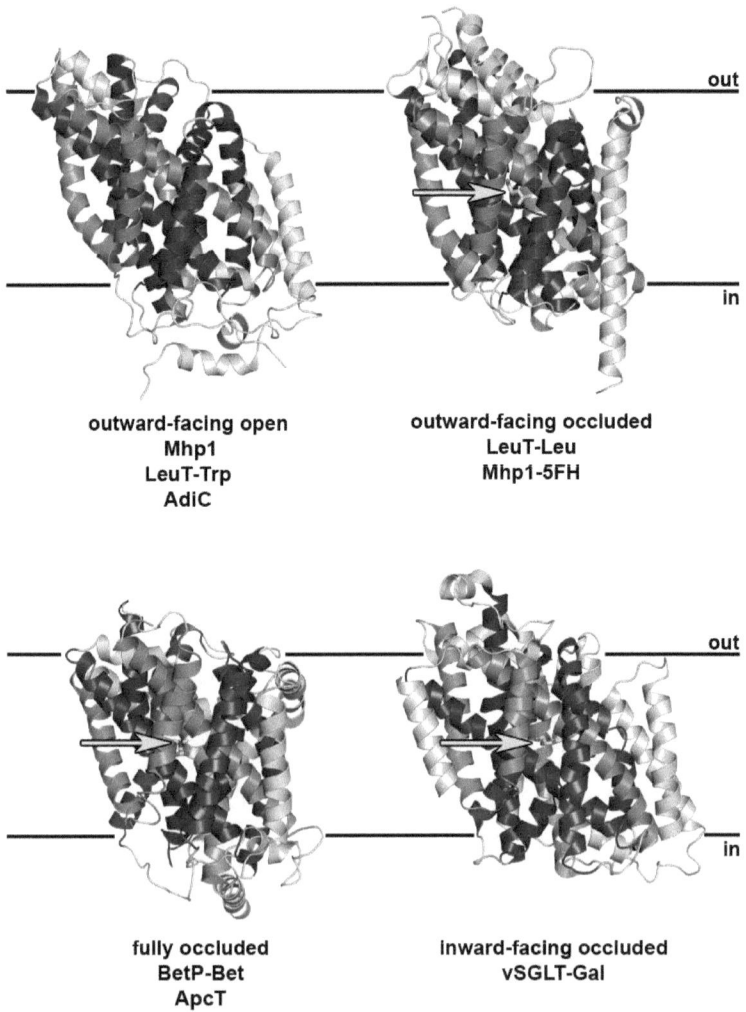

Figure 1-10 | Conformations of LeuT-type transporters

The four different conformations of six LeuT-type transporters give insights into the alternating access substrate transport mechanism. The proteins were superimposed on their inverted repeat motif and are shown parallel to the membrane. The structure of Mhp1 (PDB accession code 2JLN) represents the outward-facing open conformation. The structures of LeuT-Trp (PDB accession code 3F3A) and the substrate-free transporter AdiC (PDB accession code 3H6B and 3HOK) are also in the outward-facing open conformation. The outward-facing occluded conformation is represented by LeuT-Leu (PDB accession code 2A65) and Mhp1-5FH (PDB accession code 2JLO). In the BetP-Bet, the substrate

20

Introduction

glycine betaine (PDB accession code 2WIT) is occluded. The ApcT structure (PDB accession code 3GIA) without substrate shows the same occluded conformation as BetP. The inward-facing occluded conformation is represented by vSGLT-Gal (PDB accession code 3DH4). The substrate molecule (yellow stick representation) is bound in the center of the transporter core domain (yellow arrow).

The substrate-binding sites of the transporters are located approximately half way across the lipid membrane in the center of the protomer core (Figure 1-10; yellow arrows). The core center consists of four TM helices, the four-helix bundle, comprising the first two TM helices of each repeat. In BetP, the four-helix bundle consists of TM3 and TM4 of repeat 1 and TM8 and TM9 of repeat 2. The substrate-binding site for glycine betaine is a so-called tryptophan box consisting of three tryptophans and one tyrosine (Figure 1-11). Two of the tryptophans (Trp189 and Trp194) and the tyrosine (Tyr197) are located in TM4; the third tryptophan (Trp374) is located in TM8. The tryptophan box provides an environment in which the quaternary ammonium derivative can be bound *via* cation-π interaction and van der Waals interactions (Ressl *et al.*, 2009). The residues that are involved in glycine betaine binding in BetP are highly conserved within the BCCT family (Figure 1-8).

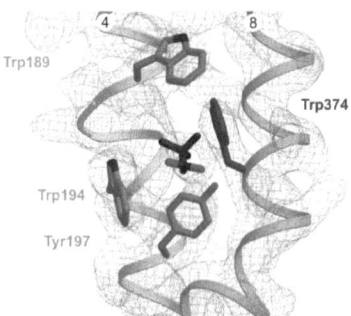

Figure 1-11 | Substrate-binding site of BetP
The residues Trp189, Trp194 and Tyr197 in TM4 and Trp374 in TM8 define the tryptophan box and provide the glycine betaine-binding site in BetP. The substrate glycine betaine is bound *via* cation-π interaction and van der Waals interaction. (Ressl *et al.*, 2009)

Introduction

1.4.1.1 Cation binding in LeuT-type transporters

In the LeuT-Leu structure two sodium sites were clearly resolved (Figure 1-12; (Yamashita *et al.*, 2005)). The two bound Na$^+$ ions have key roles in stabilizing the central core of the protein (Yamashita *et al.*, 2005). The carboxyl oxygen of the substrate leucine, residues in TM1 and residues in TM6 coordinate Na1. Interestingly, the coordination of Na1 by residues in TM1 stabilizes the unwound structure of this helix, which contains residues that are directly involved in substrate binding. In fact, all LeuT-type transporter structures reported so far show this unwound region of the first TM of repeat 1. Na2 in LeuT is located between TM1 and TM8 (Figure 1-12) and coordinated by residues in the unwound region of TM1 and in TM8.

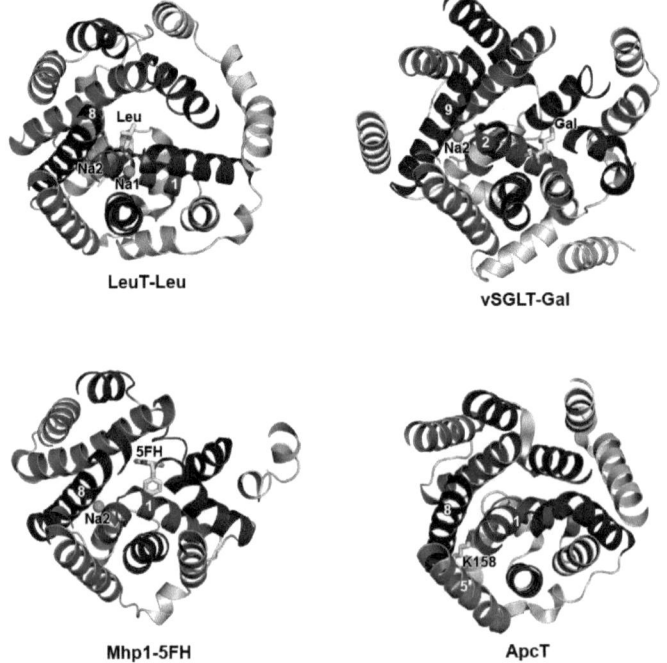

Figure 1-12 | Cation positions in LeuT-type transporters

In the LeuT-Leu structure two Na$^+$ ions have been localized. Na1 coordinates the leucine substrate and stabilizes the unwound region of TM1. In addition, Na2 stabilizes the unwound region of TM1 from

Introduction

the opposite direction. In vSGLT-Gal, one Na^+ ion was identified which superimposes on the Na2 position in LeuT-Leu. In the structure of Mhp1-5FH one Na^+ ion was modeled between TM1 and TM8, which again corresponds to the Na2 position in LeuT-Leu. The corresponding Na2 position of LeuT-Leu is occupied by a lysine (K158, TM5) in ApcT. The four proteins were superimposed on their inverted repeat motif and are shown from the periplasmic site.

Functional studies of the Na^+-dependent symporter BetP have shown that two Na^+ ions are necessary to transport one glycine betaine molecule across the membrane (Farwick *et al.*, 1995). The limited resolution of the BetP structure prevents the definite assignment of Na^+ ions. Nevertheless, the authors propose two potential Na^+ binding sites (Figure 1-13; (Ressl *et al.*, 2009)), which were structurally modeled from the Na^+ binding sites of LeuT-Leu (Yamashita *et al.*, 2005). In BetP, the carboxyl group of the substrate and residues in TM3 coordinate the first Na^+ ion (Na1), while residues in TM3 and TM7 coordinate the second Na^+ ion (Na2).

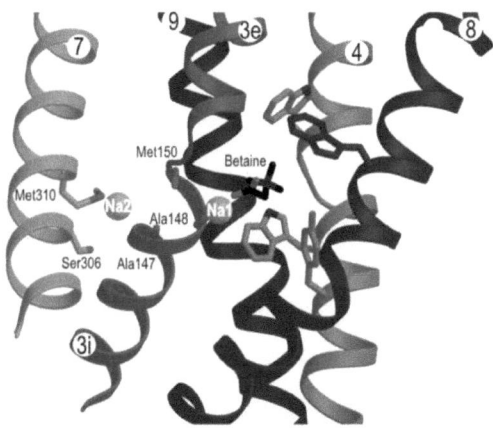

Figure 1-13 | Proposed sodium binding in BetP
The two Na^+ ions were modeled on the basis of structural alignment of BetP on the LeuT-Leu structure. The first Na^+ ion (Na1) is coordinated by the substrate glycine betaine, together with residues in TM3. The second Na^+ ion (Na2) is coordinated by residues in TM3 and TM7. (Ressl *et al.*, 2009)

In the sodium dependent galactose transporter vSGLT the one bound Na^+ ion is located at the Na2 position in the LeuT-Leu structure (Figure 1-12; (Faham et al., 2008)). This Na^+ ion is coordinated by residues in the unwound region of TM2 (TM1 in LeuT, TM3 in BetP), which also provides one residue for binding the substrate galactose. As in vSGLT, the electron density from the Na^+-coupled benzyl-hydantoin transporter Mhp1 indicates only one possible Na^+ binding site. The authors modeled this Na^+ ion between TM1 and TM8 (Figure 1-12; (Weyand et al., 2008)) which again corresponds to the Na2 position in LeuT. In Mhp1, the unwound region of TM1 provides one residue for substrate binding.

In the proton coupled amino acid transporter ApcT, the ε-amino group of a lysine (K158 in TM5), which superimposes on the LeuT-Leu Na2 position (Figure 1-12), forms a hydrogen bond to the main chain carbonyl oxygen of a glycine that is located in the unwound region of TM1 (Shaffer et al., 2009).

In some LeuT-type transporters, Na1 is required for coordinating the substrate as seen in LeuT-Leu and as proposed for BetP-Bet. In addition, Na1 stabilizes the binding environment in these transporters (unwound region of TM1 in LeuT-Leu) when the substrate is present. A positive charge at the position of Na2 probably has a regulatory function as demonstrated for ApcT, where substrate transport is strongly enhanced when lysine (K158) is protonated (Shaffer et al., 2009).

1.5 Aims of this work

Most members of the BCCT family are either Na^+ or H^+-dependent substrate transporters that are activated under osmotic stress condition. The L-carnitine/γ-butyrobetaine antiporter CaiT is an exception, as it is independent of an electrochemical gradient and becomes active under redoxstress.

The recently solved structure of the Na^+/glycine betaine permease BetP, a member of the BCCT family and close relative to CaiT, gave insights into the Na^+-dependent substrate transport mechanism. The BetP structure revealed an internal 5-TM helix motif in the transporter core. This structural feature has been reported previously for secondary transporters of different gene families. This structural homology suggests that the different gene families have a common ancestor, even though there is no significant sequence similarity. From the different conformational states shown by these transporter structures a general transport mechanism, though not yet complete, is beginning to emerge.

The 3D structure and the mechanism of Na^+-independent substrate transport in CaiT are unknown. The aims of this work are to solve the 3D X-ray structure of CaiT and to understand the substrate/product exchange mechanism.

The X-ray structure should reveal why CaiT is Na^+-independent whereas its close relative BetP is Na^+-dependent. In combination with functional studies, this should help to explain the substrate antiport mechanism in CaiT.

In addition, the CaiT structure should give further information to the general substrate transport mechanism of secondary transporters.

2 Materials and Methods

2.1 Materials

2.1.1 Instruments

Minifors Tischbioreaktor (Fermenter, INFORS AG)
Cary 50 UV-Vis Sprectrometer (Varian)
Ettan LC system (GE Healthcare)
F-4500 Flourescence Spectrophotometer (Hitachi)
TRI-CARB 1500 scintillation counter (Canberra-Packard)
Mosquito^{TM} pipetting robot (TTPLabtech)
MacroMax^{TM}-007HF diffractometer
R-AXIS IV^{++} detector (RIGAKU Americas)
FR-E^{+} SuperBright diffractometer
Saturn 944^{+} detector (RIGAKU Americas)

2.1.2 Chemicals

Chemicals used in this work were purchased from Anatrace, Avanti, Biomol, Bio-Rad, Fermentas, Gerbu, Glycon, Merck, New England Biolabs, Roth, Pierce, Roche and Sigma. γ-butyrobetaine was a kind gift from Lonza Group Ltd..

2.1.3 Reagent kits

THROMBIN CleanCleave^{TM} Kit (Sigma)

PCR Purification Kit (Qiagen)
QIAquick Gel Extraction Kit (Qiagen)
QIAprep Spin Miniprep Kit (Qiagen)
QuikChange® Site-Directed Mutagenesis Kit (Stratagene)

2.1.4 Column materials

Chelating Sepharose™ (GE Healthcare)
PD-10 Sehadex G-25M (GE Healthcare)
Superose™ 6 3.2/30 (GE Healthcare)

2.1.5 Media and antibiotics

2.1.5.1 LB-Medium

Bacto-tryptone	10 g/l
Bacto-yeast extract	5 g/l
NaCl	10 g/l

2.1.5.2 2 × YT Medium (with 5 M NaOH to pH 7 adjusted)

Bacto-tryptone	16 g/l
Bacto-yeast extract	10 g/l
NaCl	5 g/l

2.1.5.3 TB-Medium

Bacto-tryptone	12 g/l
Bacto-yeast extract	24 g/l
Glycerol	4 ml/l

Materials and Methods

Phosphate buffer   100 mM/l

2.1.5.4 *SelenoMet* Medium (Molecular Dimensions)

SelenoMet™ Medium Base   21.6 g/l
SelenoMet™ Nutrient Mix   5.1 g/l

2.1.5.5 Antibiotics

Ampicillin   100 mg/ml
Carbenicillin   100 mg/ml
Chloramphenicol   50 mg/ml (in ethanol)

2.1.6 E. coli strains

BL21(DE3) pLysS (Novagene)
Genotype: *E. coli* B F$^-$ *ompT hsdS*B(rB$^-$ mB$^-$) *gal dcm* (DE3) pLysS (Cam^R)

B834(DE3) (Novagene)
Genotype: *E. coli* B F$^-$ *ompT hsdS*B(rB$^-$ mB$^-$) *gal dcm met* (DE3)

BL21(DE3)-RIL-X (Stratagene)
Genotype: *E. coli* B F$^-$ *ompT hsdS*(rB$^-$ mB$^-$) *dcm$^+$* Tetr *gal* λ(DE3) *endA metA::Tn5*(kan^R)Hte [*argU ileY leuW* Cam^R]

2.1.7 Oligonucleotide primers

Table 2-1 | Primer for the Pm*caiT* construct

Label	Sequence	Description
CaiTProtsNdeI	5'-GGCGGCCATATGAGCAAAGATAATAAA AAGGCAGG-3'	Sense primer with *Nde*I restriction site
CaiTProtasXhoI	5'-CCGCCGCTCGAGTCAATCTTTCCAGTG ACTTTGGC-3'	Anti-sense with *Xho*I restriction site
PmCaiT_E111A_s	5'-GCTGTTTTGGGGCTCAATTGCAATATAC TACTACATTTCAAG-3'	Sense primer to mutate the glutamate at position 111 to alanine
PmCaiT_E111A_as	5'-CTTGAAATGTAGTAGTATATTGCAATTGA GCCCCAAAACAGC-3'	Anti-sense primer to mutate the glutamate at position 111 to alanine
PmCaiT_H141A_s	5'-GCTTACAGCTTATTCGCCTGGGGCCCGCT ACCTTGG-3'	Sense primer to mutate the histidine at position 111 to alanine
PmCaiT_H141A_as	5'-CCAAGGTAGCGGGCCCCAGGCGAATAAG CTGTAAGC-3'	Anti-sense primer to mutate the histidine at position 111 to alanine
ProCai_W142As	5'-GCTTACAGCTTATTCCACGCGGGCCCG CTACCTTGGGC-3'	Sense primer to mutate the tryptophan at position 142 to alanine
ProCai_W142Aas	5'-GCCCAAGGTAGCGGGCCCGCGTGGAA TAAGCTGTAAGC-3'	Anti-sense primer to mutate the tryptophan at position 142 to alanine
ProCai_W147As	5'-CTGGGGCCCGCTACCTGCGGCAACC TATAGTTTCCT-3'	Sense primer to mutate the tryptophan at position 147 to alanine
ProCai_W147Aas	5'-AGGAAACTATAGGTTGCCGCAGGTAG CGGGCCCCAG-3'	Anti-sense primer to mutate the tryptophan at position 147 to alanine
ProCaiY150As	5'-CGCTACCTTGGGCAACCGCTAGTTTC CTGTCTGTCGCC-3'	Sense primer to mutate the tyrosine at position 150 to alanine

Materials and Methods

ProCaiY150Aas	5'-GGCGACAGACAGGAAACTAGCGGTTGCCCAAGGTAGCG-3'	Anti-sense primer to mutate the tyrosine at position 150 to alanine
PCaiTR262Es	5'-GGGTAAAAATCGCCAGTGATGTGGAAACTTACCTGAGCTTC-3'	Sense primer to mutate the arginine at position 262 to glutamate
PCaiTR262Eas	5'-GAAGCTCAGGTAAGTTTCCACATCACTGGGGATTTTTACCC-3'	Anti-sense primer to mutate the arginine at position 262 to glutamate
PCaiT_W316As	5'-GGGGGTTTTCCTCAGGCTGCGACTGTCTTCTATTGGGC-3'	Sense primer to mutate the tryptophan at position 316 to alanine
PCaiT_W316Aas	5'-GCCCAATAGAAGACAGTCGCAGCCTGAGGAAAACCCCC-3'	Anti-sense primer to mutate the tryptophan at position 316 to alanine
ProtCai_W323As	5'-CTGTCTTCTATTGGGCTGCGTGGGTTATTTACGCC-3'	Sense primer to mutate the tryptophan at position 323 to alanine
ProtCai_W323Aas	5'-GGCGTAAATAACCCACGCAGCCCAATAGAAGACAG-3'	Anti-sense primer to mutate the tryptophan at position 323 to alanine
ProCai_W324As	5'-GTCTTCTATTGGGCTTGGGCGGTTATTACGCCATTC-3'	Sense primer to mutate the tryptophan at position 324 to alanine
ProCai_W324Aas	5'-GAATGGCGTAAATAACCGCCCAAGCCCAATAGAAGAC-3'	Anti-sense primer to mutate the tryptophan at position 324 to alanine
ProCaiW323324As	5'-GGACTGTCTTCTATTGGGCTGCGGCGGTTATTTACGCCATTCAAA-3'	Sense primer to mutate both the tryptophan at position 323 and the tryptophan at position 324 to alanine
ProCaiW323324Aas	5'-CTCATTTGAATGGCGTAAATAACCGCCGCAGCCCAATAGAAGACA-3'	Anti-sense primer to mutate both the tryptophan at position 323 and the tryptophan at position 324 to alanine

Materials and Methods

Label	Sequence	Description
ProCaiW324323As2	5'-CTGTCTTCTATTGGGCTGCGGCGGTT ATTTACGCC-3'	Sense primer to mutate both the tryptophan at position 323 and the tryptophan at position 324 to alanine
ProCaiW324323Aas	5'-GGCGTAAATAACCGCCGCAGCCCA ATAGAAGACAG-3'	Anti-sense primer to mutate both the tryptophan at position 323 and the tryptophan at position 324 to alanine
PCaiT_Y327As	5'-GGGCTTGGTGGGTTATTGCCGCCATT CAAATGAGTATC-3'	Sense primer to mutate the tyrosine at position 327 to alanine
PCaiT_Y327Aas	5'-GATACTCATTTGAATGGCGGCAATAAC CCACCAAGCCC-3'	Anti-sense primer to mutate the tyrosine at position 327 to alanine
ProCaiT_M331As	5'-GGTTATTTACGCCATTCAAGCGAGTATC TTCTTAGCGCG-3'	Sense primer to mutate the methionine at position 331 to alanine
ProCaiT_M331Aas	5'-CGCGCTAAGAAGATACTCGCTTGAATGGC GTAAATAACC-3'	Anti-sense primer to mutate the methionine at position 331 to alanine
ProCaiT_M331Vs	5'-GGTTATTTACGCCATTCAAGTCAGTATC TTCTTAGCGCG-3'	Sense primer to mutate the methionine at position 331 to valine
ProCaiT_M331Vas	5'-CGCGCTAAGAAGATACTGACTTGAATGGC GTAAATAACC-3'	Anti-sense primer to mutate the methionine at position 331 to valine

Table 2-2 | Primer for the Ec*caiT* construct

Label	Sequence	Description
caiT_5BamHI	5'-GCGGGATCCATGAAGAATGAAAAG AGAAAAACGGG-3'	Sense primer with *Bam*HI restriction site
caiT_3HindIII	5'-GCGTTCGAATTAATCTTTCCAGTT CTGTTTCGC-3'	Anti-sense primer with *Hind*III restriction site
caiT_5NdeI	5'-GCGCATATGATGAAGAATGAAAAG AGAAAAACGGG-3'	Sense primer with *Nde*I restriction site
caiT_3XhoI	5'-GCGCTCGAGTTAATCTTTCCAGTT CTGTTTCGC-3'	Anti-sense primer with *Xho*I restriction site

Table 2-3 | Primer for the StcaiT construct

Label	Sequence	Description
StCaiT5Nd	5'-GGCGGCCATATGAAAAATGAAAAG AAAAAATCGGG-3'	Sense primer with *Nde*I restriction site
StCaiT3Xh	5'-GGCGGCCTCGAGTTATCACTTGTC TTTCCAGTGCACTTTGGCG-3'	Anti-sense primer with *Xho*I restriction site

2.1.8 Mutants

Table 2-4 | Mutants of PmCaiT

Number	Mutant
M1	W323A
M2	W142A
M3	W324A
M4	W147A
M5	Y150A
---	---
M7	W142A/W323A
M8	W323A/W324A
M9	W142A/W323A/W324A
M10	W316A
M11	Y327A
M12	M331A
M13	M331V
M14	H141A
M15	E111A
M16	E111A/H141A
M17	R262E

All mutants were produced in this work.

2.1.9 Crystallization screens

Index^{TM} (Hampton)
Crystal Screen^{TM} (Hampton)
Crystal Screen II^{TM} (Hampton)
Detergent Screening Kits^{TM} 1 – 3 (Hampton)
Additive Screen^{TM} (Hampton)
Heavy Atom Screen Pt^{TM} (Hampton)
Aeavy Atom Screen Au^{TM} (Hampton)
Heavy Atom Screen Hg^{TM} (Hampton)
Heavy Atom Screen M1^{TM} (Hampton)
Heavy Atom Screen TM2^{TM} (Hampton)
MbClass^{TM} (Qiagen)
MbClass II^{TM} (Qiagen)
MemStart^{TM} (Jena Bioscience)
MemSyst^{TM} (Jena Bioscience)
Tantalum Cluster Derivatization Kit^{TM} (Jena Bioscience)

2.2 Molecular biological methods

2.2.1 Polymerase chain reaction

The polymerase chain reaction (PCR) is a method for *in vitro* amplification of template DNA (Mullis and Faloona, 1987) with known sequences at both ends of the DNA fragment (Saiki *et al.*, 1988). The known sequences at the ends of the DNA fragment are used to design two synthetic DNA oligonucleotides, which are complementary to the sequences of the DNA fragment. These oligonucleotides serve as primers for the *in vitro* DNA synthesis. PCR is a cyclic repetitive three-step reaction consisting of denaturation, hybridization and elongation. During the denaturation step both strands of the DNA are separated which allows the primers in the second step to hybridize to complementary sequences in the DNA strands. During the last step a thermo-stable DNA polymerase elongates the DNA primers at the 3'-OH end.

The thermo-stable *Taq*-Polymerase (Fermentas) was used for analytical PCR to create mutants of the wild-type *caiT* gene while for preparative PCR *Pfu*-Polymerase (Fermentas) was used. The *Pfu*-Polymerase possesses a 3' – 5' exonuclease activity, which corrects mutations that were produced during amplification.

Table 2-5 | PCR mixture

Component	Volume (µl)
Pfu-Polymerase buffer (10×) (Fermentas)	5
DNA template (100 – 150 ng/µl)	3
sense primer (10 mM)	5
anti-sense primer (10 mM)	3
dNTP mix (each 2 mM)	3
Pfu-Polymerase (Fermentas)	1
H_2O (autoclaved)	28

Materials and Methods

Table 2-6 | Standard PCR temperature program

Step	Temperature (°C)	Time (sec)	Cycle
Denaturation	96	180	1
Denaturation	96	60	25 – 30
Hybridization	T_{hybr}	30	25 – 30
Elongation	72	x	25 – 30
End elongation	72	300	1
	4	∞	1

The hybridization temperature depends on the nucleotide composition of the primers used in the reaction. For primers with a length of 25 nucleotides the hybridization temperature can be calculated with the following equation:

$T_{hybr} = 4 \times$ (number of G or C) $+ 2 \times$ (number of A or T)

  G = guanine   A = adenine
  C = cytosine   T = thymine

The elongation time x depends on the polymerase activity and on the DNA fragment length to be amplified. The *Pfu*-polymerase exhibits an elongation activity of 500 base pairs per minute.

2.2.2 Site-directed mutagenesis

Site-directed mutagenesis was performed with the QuikChange Site Directed Mutagenesis Kit (Stratagene) according to the manufacturer's instructions. All constructs were verified by nucleotide sequencing.

2.2.3 DNA cleavage using restriction endonucleases

Hydrolytic cleavage of double-stranded DNA was performed using restriction endonucleases from New England Biolabs according to the manufacturer's instructions. Hydrolysis with two different enzymes was set up in the same test tube as long as the reaction conditions for both enzymes were similar. The reactions were performed in succession when the reaction conditions for both endonuclease enzymes were distinctly different.

Cleaved DNA fragments were purified either with the PCR Purification Kit (Qiagen) or by agarose gel electrophoresis. DNA fragments were extracted with the QIAquick Gel Extraction Kit (Qiagen) according to the manufacturer's instructions.

2.2.4 Agarose gel electrophoresis

Depending on the length of the DNA fragments a separation was achieved by a 1 – 1.5 % (w/v) agarose gel in 1× TAE buffer (Table 2-7). The DNA samples were mixed with 6× DNA sample buffer (Table 2-8) before they were loaded onto the gel. The electrophoresis was performed at 100 V (distance of electrodes 10 cm).

For visualization of the DNA fragments ethidium bromide (0.5 µg/ml) was added to the agarose gels. Ethidium bromide is an organic fluorescent dye with aromatic rings that intercalates with the nucleotides of the DNA. The fluorescent dye can be illuminated at the excitation wavelength $\lambda = 312$ nm.

Table 2-7 | TEA buffer

TAE buffer (1×)	
Tris-acetate (pH 8.3)	40 mM
EDTA	1 mM

Table 2-8 | DNA sample buffer

DNA sample buffer (6×)	
bromophenol blue	0.25 % (w/v)
sucrose	40 % (w/v)

2.2.5 DNA concentration and purity

The concentration and purity of DNA is determined photometrically. The concentration of double-stranded DNA can be calculated with the optical density (OD) at the absorption maximum at 260 nm (OD_{260}). The reference value is OD_{260} = 1 for a pure 50 µg/ml DNA solution.

The purity of a DNA solution can be estimated by the OD_{260} to OD_{280} ratio (Glasel, 1995; Warburg and Christian, 1942). An absorption at wavelength $\lambda = 280$ indicates the presence of proteins, which were co-purified during DNA purification. A pure DNA solution shows an OD_{260} to OD_{280} ratio of 1.8 – 2.0.

2.2.6 Ligation of DNA fragments

For a ligation reaction, 50 – 100 ng of purified vector DNA and a 3 – 5 fold molar excess of purified insert DNA was mixed with T4-ligase buffer (10×) and T4-DNA-ligase (200 U). Both the vector and the insert DNA have been cleaved by the same restriction enzymes before being added to the ligation reaction. The total volume of the reaction solution was 20 µl. The ligation was incubated overnight at 8 °C.

The *caiT* constructs used in this work were cloned into a pET-15b (Novagen) vector using the restriction nucleases *Nde*I (NEB) and *Xho*I (NEB).

2.2.7 Preparation and transformation of chemically competent cells

Chemically competent *E. coli* cells were produced by the Chung method (Chung *et al.*, 1989). This method uses TSS medium (Table 2-9), which is based on LB medium.

An *E. coli* strain to be made competent was grown in 100 ml LB medium overnight (37 °C, 125 rpm). The next morning, the culture was adjusted to an OD_{600} of 0.1 and cultured (20 °C, 125 rpm) again to an early exponential phase (OD_{600} = 0.4). To cool down the culture slowly it was incubated for 30 minutes at 4 °C before the cells were pelleted (4 °C, 2500 g, 10 min). The supernatant was removed and the cells were resuspended in TSS medium with 1/10 of the original culture volume. The cell suspension was aliquoted (100 – 200 µl) and shock frozen in liquid nitrogen. The aliquots were stored at -80 °C.

Table 2-9 | TSS medium

TSS medium	
LB-medium (without NaOH)	
+ PEG 8000 (w/v)	10 %
+ DMSO (v/v)	5 %
+ $MgCl_2 \times 6\ H_2O$ (w/v)	50 mM
+ Glycerol (v/v)	15 %
Sterilized by filtration (exclusion size $\varnothing$ 0.45 µm)	

The transformation reaction was set up with 100 – 200 µl of chemically competent *E. coli* cells and either 10 µl of ligation reaction solution or 25 – 50 ng of circular vector DNA. After addition of DNA, the competent cells were incubated for 30 minutes at 4 °C. To increase the transformation rate a heat shock at 42 °C for 90 sec was administered. The cells were again incubated at 4 °C for 5 min before 1 ml of LB medium was added and the culture incubated at 37 °C for 45 minutes (125 rpm). This incubation step is necessary to allow the expression of the antibiotic resistance gene located on the inserted vector. After incubation, the cells were plated on LB-agar-plates with the corresponding selection marker (e.g. ampicillin) and incubated at 37 °C overnight.

2.2.8 Isolation of vector DNA

Transformed cells were grown overnight at 37 °C in 5 ml LB medium with the appropriate selection marker. The next morning, cells were harvested by centrifugation (20 °C, 2500 g, 5 min) and vector DNA was isolated using the QIAprep Spin Miniprep Kit (Qiagen) according to the manufacturer's instructions.

2.2.9 Preparation of bacteria glycerol stocks

A single isolated *E. coli* colony, grown on a LB-agar-plate under selection conditions, was transferred into 2 ml LB medium containing the selection marker. The culture was grown overnight (37 °C, 125 rpm) and 30 % (v/v) glycerol was added to the culture before it was shock frozen in liquid nitrogen. The bacteria glycerol stocks were stored at -80 °C.

2.3 Biochemical methods

2.3.1 Protein production of native CaiT

The genes encoding EcCaiT, PmCaiT and StCaiT were cloned into the expression vector pET-15b (Novagen) using the restriction enzymes *Nde*I (NEB) and *Xho*I (NEB). In the pET-15b vector the target gene is under control of a strong bacteriophage T7 transcription promoter. The plasmids were transferred into the expression host *E. coli* BL21(DE3) pLysS.

BL21(DE3) cells are lysogens of bacteriophage DE3, a lambda derivative that has the immunity region of phage 21 and carries a chromosomal copy of the T7 RNA polymerase under control of the *lacUV5* promotor. The *lacUV5* promotor is inducible by IPTG. ITPG induction activates the *lacUV5* promotor, which induces the transcription of T7 RNA polymerase, which in turn leads to expression of the recombinant protein.

BL21(DE3) pLysS cells contain an additional compatible plasmid with a selectivity marker that provides a small amount of T7 lysozyme, a natural inhibitor of the T7 RNA polymerase (Moffatt and Studier, 1987; Studier, 1991). T7 lysozyme inhibits expression of the target genes before ITPG induction and is also able to cut a specific bond in the peptidoglycan layer of the *E. coli* cell wall. This allows the cells to be lysed simply by freezing and thawing.

For protein production 50 µl of a thawed glycerol stock containing transformed BL21(DE3) pLysS cells were used to inoculate 100 ml of LB medium containing 100 µg/ml ampicillin and 50 µg/ml chloramphenicol. After overnight growth at 37 °C with shaking (150 rpm), 10 ml of the culture were used to inoculate 2 l of 2×YT medium supplemented with the appropriate antibiotics. This culture was grown at 37 °C with shaking (120 rpm) until an OD_{600} of 0.6 – 0.9 was reached. To start over-production of recombinant CaiT, IPTG was added to a final concentration of 0.5 mM. After over-production for 4 h at 30 °C with shaking (120 rpm), cells were harvested by centrifugation (4 °C, 4000 g, 20 min), with a potter homogenized in buffer containing 25 mM HEPES (pH 7.5), 100 mM NaCl and 0.1 mM PefaBloc (Biomol), and frozen overnight at -20 °C.

2.3.2 Protein production of selenomethionine-labeled protein

The pET-15b vector containing the genes encoding either EcCaiT or PmCaiT was transferred into the expression hosts B834(DE3) (Novagene) or BL21(DE3)-RIL-X (Stratagene). Both cell strains are auxotrophic for methionine and useful for selenomethionine (SeMet)-labeling of proteins used for protein crystallography. BL21(DE3)-RIL-X cells contain an additional compatible plasmid with a second selectivity marker (chloramphenicol) and extra copies of *argU*, *ileY* and *leuW* tRNA genes. These genes encode tRNAs that recognize certain condons for arginine (R), isoleucin (I) and leucin (L), which are rarely used in *E. coli*. However, these genes were not used for expression of SeMet-labeled CaiT.

CaiT was SeMet-labeled by different methods. The first, flask based method, was very similar to the method used for unlabeled recombinant protein production, while the second approach was a fed-batch fermentation strategy to achieve a higher yield of SeMet-labeled protein.

2.3.2.1 Flask based production of SeMet-labeled protein

For flask based SeMet-labeled protein production, 50 µl of a thawed glycerol stock containing either transformed B834(DE3) cells or BL21(DE3)-RIL-X cells were used to inoculate 100 ml of LB medium containing 100 µg/ml ampicillin. After overnight growth at 37 °C with shaking (150 rpm), the cells were pelleted by centrifugation (20 °C, 4000 g, 20 min). The supernatant was removed and the cells were washed and resuspended in 40 ml *SelenoMet* medium (Molecular Dimensions) supplemented with the appropriate antibiotic and 80 mg/l methionine. The complete

Materials and Methods

40 ml cell suspension was used to inoculate 1 l of *SelenoMet* medium supplemented with the appropriate selection marker and 80 mg/l methionine. This culture was again grown overnight at 37 °C with shaking (125 rpm) before 100 ml were used to inoculate 2 l of *SelenoMet* medium supplemented with 100 mg/l ampicillin and 40 mg/l SeMet. The culture was grown at 37 °C with shaking (120 rpm) until an OD_{600} of 0.6 – 0.9 was reached. Over-production of recombinant SeMet-labeled CaiT was induced with IPTG. The final concentration of ITPG in each culture was 0.5 mM. After over-production for 6 – 8 h at 30 °C, cells were harvested by centrifugation (4 °C, 4000 g, 20 min), homogenized in buffer containing 25 mM HEPES (pH 7.5), 100 mM NaCl, 2 mM TCEP and 0.1 mM PefaBloc (Biomol), and frozen overnight at -20 °C.

2.3.2.2 Fed-batch fermentation

For fed-batch fermentation, 50 µl of a thawed glycerol stock containing either transformed B834(DE3) cells or BL21(DE3)-RIL-X cells were used to inoculate 100 ml of TB medium with 100 µg/ml ampicillin. The cells were cultured overnight at 37 °C with shaking (150 rpm), pelleted by centrifugation (20 °C, 4000 g, 20 min), washed and resuspended in 40 ml *SelenoMet* medium supplemented with the appropriate antibiotic and 80 µg/ml methionine. The complete 40 ml cell suspension was used to inoculate 3 l of *SelenoMet* medium supplemented with the 100 mg/l ampicillin and 80 mg/l methionine. The cells were grown at 37 °C and 200 rpm until the concentration of dissolved O_2 in the medium decreased (Figure 2-1). This signaled a depletion of methionine. At this point the culture showed an OD_{600} of 1.3 – 1.7 and the feeding with SeMet was started to a final concentration of 80 mg/l. SeMet-labeled protein production was induced 15 – 30 minutes after starting the SeMet feed by adding IPTG to a final concentration of 0.5 mM. The culture continued to grow to a final OD_{600} of 2.4 – 2.6. The cells were harvested by centrifugation (4 °C, 4000 g, 20 min), homogenized with a potter in buffer containing 25 mM HEPES (pH 7.5), 100 mM NaCl, 2 mM TCEP and 0.1 mM PefaBloc (Biomol), and frozen overnight at -20 °C.

Materials and Methods

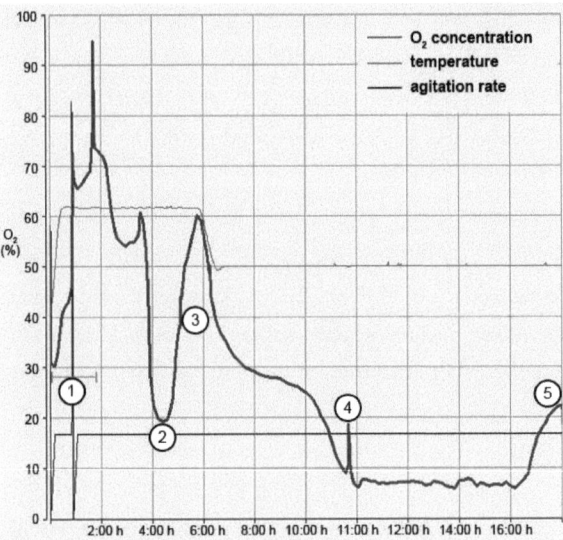

Figure 2-1 | Fed-batch fermentation diagram

The blue line indicates the concentration of dissolved O_2, the red line shows the temperature (cell growth at 37 °C, protein over-expression at 30 °C) and the black line shows the agitation rate (constant at 200 rpm). The starting phase indicated by 1 was used to calibrate the O_2 electrode. At time point 2 the culture was supplemented with SeMet, at time point 3 over-expression was induced by IPTG, at time point 4 a mix of nutrients, SeMet (final concentration 80 mg/l) and ampicillin (final concentration 100 mg/l) was added. At time point 5 the cells were harvested (usually 10 – 12 hours after induction).

2.3.3 Cell disruption

2.3.3.1 Enzymatic cell disruption

Cell were disrupted enzymatically simply by thawing BL21(DE3) pLysS cells which contain T7 lysozyme. Freezing and thawing of the cells lead to cell wall breakage of a few cells and the release of T7 lysozyme into the solution. This in turn caused the lysis of more cells in a chain reaction manner.

Materials and Methods

Release of chromosomal *E. coli* DNA during cell lysis causes the cell suspension to become viscous. The cell suspension became fluid again after the addition of small amounts of DNAse I (Roche).

2.3.3.2 Cell disruption using a microflidizer

Cells were in addition, or as an alternative to enzymatic disruption, disrupted using a microfluidizer (Model M-110L, Microfluidics Corp., Newton, MA) with 3 passes at 60,000 psi (8.7 MPa). The microfluidizer pushes cells at high pressure through a narrow nozzle ($\varnothing$ < 0.5 mm). After leaving the nozzle the cells expand due to the release of high pressure. The disruption of the cells is caused by cell expansion together with the shearing force, which occurs when the cells leave the nozzle. To avoid a temperature increase of the cell lysate, all containers and the microfluidizer were precooled with ice.

2.3.4 Protein purification

2.3.4.1 Membrane preparation

To isolate *E. coli* membranes, two centrifugation steps were needed after cell disruption. Cell debris was removed by low-speed centrifugation (4°C, 10,000 g, 30 min), after which membranes were isolated by ultracentrifugation (4°C, 125,000 g, 90 min). The membrane pellet was resuspended in membrane buffer. Aliquots of 20 – 40 ml volume with a total protein concentration of 10 mg/ml were frozen in liquid nitrogen and kept at -80 °C.

Table 2-10 | Membrane buffer

Component	Concentration
HEPES or Tris-HCl, pH 7.5	50 mM
NaCl or KCl	200 mM
TCEP	2 – 4 mM
Glycerol	20 % (v/v)

2.3.4.2 Protein solubilization

For CaiT solubilization, membranes were thawed and incubated with constant stirring either with 1.5 – 2 % DDM for 2 hours at 4 °C or with 1.5 – 2% Cymal-5 overnight at 4 °C.

Table 2-11 | Solubilization buffer

Component	Concentration
HEPES or Tris-HCl, pH 7.5	25 mM
NaCl or KCl	100 mM
TCEP	1 – 2 mM
Glycerol	10 % (v/v)
DDM or Cymal-5	1.5 – 2 % (w/v)

After solubilization an ultracentrifugation step (4 °C, 125,000 g, 40 min) removed the unsolubilized fraction. The supernatant, which contained solubilized CaiT, was supplemented with 50 mM imidazole (final concentration) to prevent unspecific binding of contaminating proteins in the next purification step.

Materials and Methods

2.3.4.3 Immobilized Metal Ion Affinity Chromatography (IMAC)

The recombinantly expressed protein contains an N-terminal hexa-histidine tag (His_6-tag) and a thrombin cleavage site included in the linker region between the His_6-tag and CaiT. The first purification step involved a Ni^{2+} loaded Chelating Sepharose™ (GE Healthcare) column (Ni^{2+}-Sepharose column). For 40 ml of solubilized fraction, 2 ml of wet column material was used. The column material was equilibrated with the appropriate buffer (Table 2-12) before the solubilized fraction was loaded. The binding of the target protein to the column results from the formation of a chelate complex between Ni^{2+} ions, the column material, and two histidine residues of the His_6-tag. Before the protein was eluted, the column was washed twice with at least 5 column volumes. The first washing step was performed with Wash 1 buffer (Table 2-12) and the second with Wash 2 buffer (Table 2-12). Aliquots of 500 – 1000 µl were collected and fractions containing the highest concentration of protein were pooled.

Materials and Methods

Table 2-12 | **Buffers used for affinity chromatography purification**

Buffer	Component	Concentration
Equilibration / Wash 1	HEPES or Tris-HCl, pH 7.5	25 mM
	NaCl or KCl	50 – 100 mM
	TCEP	1 – 2 mM
	DDM or Cymal-5	0.02 – 0.2 % (w/v)
	Imidazole	50 mM
	Glycerol	10 % (v/v)
Wash 2	HEPES or Tris-HCl, pH 7.5	25 mM
	NaCl or KCl	50 – 100 mM
	TCEP	1 – 2 mM
	DDM or Cymal-5	0.02 – 0.2 % (w/v)
	Imidazole	50 mM
Elution	HEPES or Tris-HCl, pH 7.5	25 mM
	NaCl or KCl	50 – 100 mM
	TCEP	1 – 2 mM
	DDM or Cymal-5	0.02 – 0.2 % (w/v)
	Imidazole	200 mM

2.3.4.4 Size exclusion chromatography (SEC)

Preparative SEC was used to remove imidazole after eluting the protein from the Ni^{2+} column on a PD-10 Sephadex G-25M (GE Healthcare) column. The column was equilibrated with the appropriate buffer (Table 2-13) before max. 2.5 ml protein solution was loaded on the column. Aliquots of 200 – 500 µl were collected and fractions with detectable protein content were pooled.

Analytical SEC was carried out using the Ettan LC system (GE Healthcare) and a Superose™ 6 (3.2/30) column (GE Healthcare). The flow rate was kept constant at 50 µl/min. All buffers (Table 2-13) used during analytical SEC were filtered (exclusion size 0.2 µm) and degassed.

Materials and Methods

Table 2-13 | Buffers used for preparative and analytic SEC

Buffer	Component	Concentration
SEC	HEPES or Tris-HCl, pH 7.5	25 mM
	NaCl or KCl	50 – 100 mM
	TCEP	1 mM
	DDM or Cymal-5	0.02 – 0.2 %

2.3.4.5 Proteolytic cleavage of the fusion tag

The His_6-tag was removed using the THROMBIN ClenCleave™ Kit (Sigma). Thrombin is a protease that selectively cleaves between the arginine and glycine residues of the cleavage site (Leu-Val-Pro-Arg↓Gly-Ser) and can therefore be used effectively to remove the purification tag.

The washing and cleavage buffers were adjusted to protein compatible conditions (50 mM HEPES/Tris-HCl, pH 7.5, 50 – 100 mM NaCl/KCl, 1 mM TCEP, 10 mM $CaCl_2$, 0.02 – 0.2 % DDM/Cymal-5). The cleavage reaction was incubated at 4 °C for 3 – 4 hours with constant stirring. All other steps were performed according to the manufacturer's instructions.

2.3.4.6 Protein concentration

Protein was concentrated by centrifugation (4 °C, 4,000 g) using the Vivaspin 4 or Vivaspin 500 concentrators with a molecular weight cut-off of 100,000 kDa (Vivascience). To avoid protein aggregation due to high local protein concentration, the sample was mixed every 10 minutes.

2.3.5 Detergent concentration

The detergent concentration was analyzed after the protein purification was completed. The method makes use of the sugar moiety in the detergent which is quantified by a colorimetric reaction with phenol and sulfuric acid (Fox and Robyt, 1991; Urbani and Warne, 2005).

The reaction was set up in small glass tubes (total volume 750 µl, VWR) that were placed on ice. 25 µl of a diluted protein sample (containing 20 – 50 µg protein) was mixed with 25 µl 5 % (w/v) phenol before 150 µl of concentrated sulfuric acid was added. The reaction was mixed carefully and incubated at 80 °C for 30 minutes. After cooling to room temperature the samples were pipetted into wells of a 96-well plate (Nunc). The absorption was measured at 492 nm using the SpectraMax M2 (Molecular Devices) microplate reader. Standard curves were calculated using DDM or Cymal-5.

2.3.6 Protein concentration

2.3.6.1 Bradford assay

The protein concentration of a membrane solution was estimated by the Bradford assay (Bradford, 1976). The Bradford Reagent (Sigma) contains the Coomassie Brilliant Blue G250 dye that binds to basic amino acids like arginine, histidine, and lysine but also interacts with hydrophobic amino acids like phenylalanine, tryptophan and tyrosine. The binding of the protein to the dye stabilizes its anionic from which leads to a bathochromic shift of the absorption maximum from 465 nm (without bound protein) to 595 nm (with bound protein). The intensity of the blue colouration is directly proportional to the protein concentration. A standard curve was calculated using BSA and the absorption was measured at 595 nm.

Materials and Methods

2.3.6.2 Protein denaturing method

The protein concentration of purified protein was estimated by measuring the absorbance at 280 nm. The extinction coefficient ε and the molar mass M_r of CaiT were calculated from the amino acid sequence using the ExPASy ProtParam tool (www.expasy.ch/tools/protparam.html). Before measuring the absorption, the purified protein was denatured in 6 M guanidinium hydrochloride. Guanidinium hydrochloride was also used as the blank reference. The protein concentration was calculated using Equation 2-1.

Equation 2-1:

$$c(CaiT) = \frac{Abs.(\lambda_{280}) \bullet 100}{\varepsilon(CaiT)} \times M_r(CaiT)$$

ε (EcCaiT) = 153,945 M_r (EcCaiT) = 56,587 Da
ε (PmCaiT) = 152,915 M_r (PmCaiT) = 56,318 Da
ε (StCaiT) = 150,965 M_r (StCaiT) = 56,748 Da

2.3.6.3 Amido Black method

The concentration of protein reconstituted into liposomes was determined using the Amido Black method (Schaffner and Weissmann, 1973). The Amido Black method is a colorimetric assay for measuring total protein concentration in a given sample with a concentration down to 0.75 µg/ml. The protein is quantitatively precipitated by trichloroacetic acid (TCA) and collected on membrane filters by filtration. The protein collected on the filter can be quantitatively determined by staining with the dye amido black.

A 2 – 5 µl protein sample was placed in a test tube and water was added to a final volume of 225 µl. The solution was first mixed with 30 µl of a buffer containing

1 M Tris-HCl, pH 7.4 and 2 % (w/v) SDS before the protein was precipitated with 50 µl 90 % (v/v) TCA. The mixture was vortexed and kept 2 – 5 minutes at room temperature. A round filter (MILLIPORE, type HAWP02500, 0.45 µm), which was carefully marked with pencil dots, was washed with water and placed on a suction device. The samples were carefully dropped on to the respective pencil dots and sucked through the filter with a vacuum pump. The protein spots were directly washed with 200 µl 6 % (v/v) TCA and the whole filter was again rinsed with 1 – 2 ml 6 % (v/v) TCA. After completing all the filtrations, the filter was moved to a Petri dish and immersed for 2 – 3 minutes in Amido Black solution (Table 2-14). To remove excess solution the filter was first rinsed with water before it was washed under agitation for 2 – 3 minutes in destaining solution (Table 2-14). The spots containing protein appeared blue and were cut out and placed in a test tube. 1 ml of eluant solution (Table 2-14) was added to the test tubes and the tubes were incubated under agitation for 10 minutes at room temperature. The OD of the eluant solution was measured at 630 nm. A standard curve was calculated using BSA (concentration range: 0 µg to 10 µg).

Table 2-14 | Solutions for the Amido Black Assay

Solution	Component	Concentration
Amido Black	Amido Black	0.25 % (w/v)
	Methanol	45 % (v/v)
	Glacial acetic acid	10 % (v/v)
Destaining	Methanol	90 % (v/v)
	Glacial acetic acid	2 % (v/v)
Eluant	NaOH	25 mM
	EDTA	50 µM
	Ethanol	50 % (v/v)

2.3.7 Polyacrylamid gel electrophoresis (PAGE)

2.3.7.1 Blue-Native (BN)-PAGE

The discontinuous BN-PAGE (Schagger and von Jagow, 1991) was used to detect undenatured, solubilized proteins after purification. The method uses Coomassie Blue G250, which induces a charge shift on the proteins and stays tightly bound to it even upon high dilution and in the presence of ε-aminocaproic acid, which improves the solubility of membrane proteins.

Native gels (4 – 12 % Tris-Glycine gels) used for BN-PAGE were purchased from Invitrogen. Protein samples (3 – 7.5 µg) were pre-diluted with 10× sample buffer (Table 2-15) before loading onto the gel. Electrophoresis was started at 80 V until the protein samples were concentrated within the stacking gel. The gels were run at a constant voltage (150 V) for 4 – 5 hours. The gels were distained with 10 % (v/v) Ethanol and 10 % (v/v) Glacial acetic acid.

Table 2-15 | BN-PAGE buffers

Buffer	Component	Concentration
Anode	BisTris, pH 7.0	50 mM
Cathode	Tricin	50 mM
	BisTris, pH 7.0	15 mM
	Coomassie-Brilliant-Blue G250	0.2% (w/v)
Sample (10×)	ε-aminocaproic acid	500 mM
	BisTris, pH 7.0	150 mM
	Glycerol	10 % (v/v)
	Coomassie-Briliant-Blue G250	0.2% (w/v)

2.3.7.2 Denaturing SDS-PAGE

To analyze the purity of a protein sample a discontinuous PAGE under denaturing conditions was performed according to Laemmli (Laemmli, 1970). 15 % resolving gels with 5 % stacking gels (Table 2-16) were prepared and stored at 4 °C. Before electrophoresis the protein samples were diluted with 4× sample buffer (Table 2-18) which contains the anionic detergent SDS that denatures, linearizes, and charges proteins according to their amino acid length. Electrophoresis was started at 80 V until the protein samples entered the stacking gel and afterwards run at a constant voltage of 120 V for 45 minutes. Gels were stained with Coomassie (Table 2-19) according to the protocol by Studier (Studier, 2005).

Table 2-16 | SDS-PAGE gel mixture for two gels

Gel	Component	Volume
Resolving (15 %)	H_2O	2.3 ml
	30 % (v/v) acryl-bisacrylamide mix	5.0 ml
	1.5 M Tris-HCl, pH 6.8	2.5 ml
	10 % (w/v) SDS	100 µl
	10 % (w/v) APS (fresh)	100 µl
	TEMED	5 µl
Stacking (5 %)	H_2O	2.7 ml
	30 % (v/v) acryl-bisacrylamide mix	670 µl
	1.5 M Tris-HCl, pH 6.8	500 µl
	10 % (w/v) SDS	40 µl
	10 % (w/v) APS (fresh)	40 µl
	TEMED	5 µl

Table 2-17 | SDS-PAGE running buffer

Component	Concentration
Tris-HCL (no pH adjustment)	250 mM
Glycine	1.92 M
SDS	1 % (w/v)

Table 2-18 | 4× sample buffer

Component	Concentration
Tris-HCl (no pH adjustment)	250 mM
SDS	10 %
Glycerol	50 %
TCEP	10 mM
Bromophenol blue	0.02 % (w/v)

Table 2-19 | Coomassie staining solutions

	Component	Concentration
Solution I:	Ethanol	50 % (v/v)
	Acetic acid	10 % (v/v)
Solution II:	Ethanol	5 % (v/v)
	Acetic acid	7.5 % (v/v)
Coomassie stock solution:	Ethanol	95 % (v/v)
	Coomassie brilliant blue R250	0.25 % (w/v)

2.3.8 Western blot analysis

Transfer of proteins from SDS-PAGE gels to methanol-activated PVDF membranes (Immobilon™ -P Transfer Membrane, pore size 0.45 µm, MILLIPORE) was performed using a semi-dry system (Bio-Rad). The PVDF membrane and the gel were sandwiched between Whatman 3MM filter paper (Schleicher & Schüll), presoaked with transfer buffer (Table 2-20). To ensure complete transfer of the proteins to the PVDF membrane, blotting was carried out for 45 minutes at 15 V. After transfer, the PVDF membrane was incubated at room temperature for 1 hour in blocking buffer (Table 2-20) to cover the membrane with protein so the antibodies cannot bind to the membrane unspecifically. The membrane was subsequently washed 3 – 5 times in TBS buffer (Table 2-20).

For immuno-detection, the PVDF membrane was incubated in primary antibody (Monoclonal Anti-polyHISTIDINE clone HIS-1, Sigma; 1 : 1000 dilution) for 2 hours at room temperature or overnight at 4 °C. The membrane was washed 3 – 5 times in TBS buffer (Table 2-20) and incubated with the second antibody (Anti-Mouse IgG (whole molecule)-Alkaline Phosphatase, antibody produced in rabbit, Sigma; 1 : 5000 dilution) for 2 hours at room temperature or overnight at 4 °C. After washing 3 – 5 times in TBS (Table 2-20), the membrane was incubated in 10 ml developer buffer (SigmaFAST^{TM} BCIP/NBT, Sigma; Table 2-20) until bands appeared. To stop the reaction the PVDF membrane was washed with deionized water.

Table 2-20 | Buffers for Western blot analysis

Buffer	Component	Concentration
Transfer	Tris-HCl (no pH adjustment)	25 mM
	Glycine	192 mM
	SDS	0.1 % (w/v)
TBS (10×)	Tris-HCl, pH 7.5	500 mM
	NaCl	1.5 M
Blocking	TBS buffer	1×
	Milk powder	5 % (w/v)
Antibody	TBS	1×
	Milk powder	0.5 % (w/v)
Developer	Tris-HCl, pH 7.5	25 mM
	SigmaFAST^{TM} BCIP/NBT	1 tablet

2.3.9 Thin-layer chromatography (TLC)

TLC was performed to analyze the lipid content of solubilized membrane protein after purification. A small drop of purified protein (10 – 15 µg) in detergent solution was applied to a TLC plate (10×10 cm^2, pre-coated, Silica 60, Merck) about 1 cm from the base. The plate was developed in a closed tank containing a solvent mixture (chloroform : methanol : water = 69 % : 27 % : 4 % (v/v/v)). After drying, the TLC plate was stained with iodine vapour, which binds to carbon-carbon double bonds (C = C) of unsaturated fatty acids and therefore stains lipids non-specifically. Phospholipids were detected using the molybdenum blue spray reagent (1.3 % (w/v) molybdenum oxide (MoO$_3$) in 4.2 M sulfuric acid, 1 : 1 dilution, Sigma). Under acidic condition molybdenum oxide reacts with the phosphate group of phospholipids and forms the blue heteropoly-molybdenum anion (PMo$^V_4$Mo$^{VI}_8$O$_{40}^{7-}$).

2.3.10 Protein reconstitution into liposomes

For reconstitution of CaiT protein into liposomes, *E. coli* polar lipids (*E. coli* Polar Lipid Extract, Avanti) were used. An *E. coli* polar lipids (EPL) aliquot in chloroform : methanol was dried under nitrogen and resuspended in KP$_i$ buffer (50 mM KP$_i$, pH 7.5, 1 mM TCEP) to a final concentration of 20 mg/ml. Aliquots of 800 µl were frozen in liquid nitrogen and stored at -80 °C.

The lipid solution was slowly thawed at room temperature before it was extruded 15 – 20 times through a 400 nm membrane filter (Polycarbonate Membrane, pore size = 400 nm, Avestin) and diluted to a final concentration of 5 mg/ml. The liposomes were titrated with Triton X100 (10 % stock solution (w/v)) to the onset of solubilization, which was detected by measuring the absorbance at 540 nm. Protein was added at a lipid-to-protein ratio of 20 : 1 (w/w). The protein-liposome solution was incubated for 20 minutes at room temperature under gentle agitation before detergent was removed by adding Bio-Beads (SM-2 Macroporous Beads, Bio-Rad). Proteoliposomes were collected by ultra-centrifugation (20 °C, 100,000 g, 35 min), washed and resuspended in KP$_i$ buffer to a final lipid concentration of 60 mg/ml.

Materials and Methods

Aliquots of 50 µl were frozen in liquid nitrogen and stored at -80 °C. Reconstitution was checked by freeze-fracture electron microscopy.

2.3.11 Transport measurements

Substrate exchange was measured by recording the uptake of L-[N-Methyl-^{14}C]-carnitine hydrochloride (^{14}C-L-carnitine, Perkin Elmer) into CaiT-containing liposomes preloaded with a saturating concentration of non-radioactive substrate. Usually 10 mM of potential substrate such as glycine (Sigma), glycine betaine (Sigma), γ-butyrobetaine (a kind gift from the Lonza Group Ltd.) or L-carnitine (Sigma) was used. Proteoliposomes were loaded by two rounds of freeze-thawing in the presence of substrate. After the second freeze-thaw cycle the proteoliposomes were extruded first through a 200 nm membrane filter (Polycarbonate Membrane, pore diameter 200 nm, Avestin) followed by extrusion through a 400 nm membrane filter (Polycarbonate Membrane, pore diameter 400 nm, Avestin). The extraliposomal substrate was removed by washing the proteoliposomes with KP_i buffer. Proteoliposomes were collected by ultra-centrifugation (20°C, 100,000 g, 35 min). Substrate uptake was initiated by aliquots (2 µl) of the proteoliposome suspension diluted into 400 µl KP_i buffer with different concentration of ^{14}C-L-carnitine (typically 6 to 620 µM, 2.5 µCi ml^{-1}). At time points of 5 to 30 sec, ^{14}C-L-carnitine incorporated into the proteoliposomes was determined with a scintillation counter TRI-CARB 1500 (Canberra-Packard). All transport assays were repeated three times.

2.4 Biophysical methods

2.4.1 Fluorescence

Fluorescence is the spontaneous emitted radiation by a substance that has been excited by absorption of radiation at a different, usually shorter, wavelength. Fluorescence typically occurs from aromatic molecules, which have rigid and inflexible structures (e.g. flourescein or tryptophan). The Jablonski diagram (Figure 2-2; (Jablonski, 1935)) illustrates the processes that occur between the absorption and emission of light. The initial absorption of a photon transfers the fluorophore to an excited electronic state by promoting an electron from a lower to a higher energy state. Some of the initially absorbed energy is lost in transitions between vibrational energy levels (non-radiative transition). Radiative transition or fluorescence occurs when the molecule returns from the excited electronic state (S_1) to the ground state (S_0).

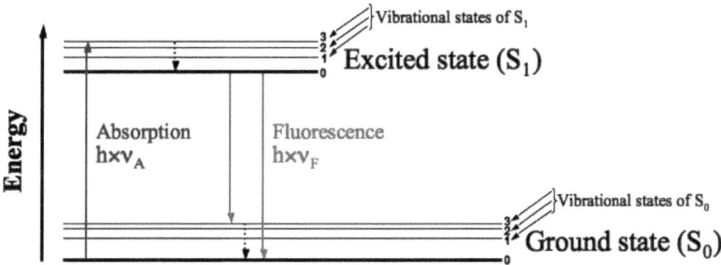

Figure 2-2 | Physical basis of fluorescence
Simplified Jablonski diagram (Jablonski, 1935). The singlet ground (S_0) and the first electronic (S_1) states are shown. At each of the two different electronic energy levels, the fluorophore can drop to different vibrational energy levels (0, 1, 2, 3, …). Absorption of a photon by a fluorophore corresponds to a transition from S_0 to S_1. If the fluorophore has a comparable rigid structure with few vibrational energy levels, fluorescence may occur. Straight lines indicate radiative transition (absorption or emission) of a photon, while dashed lines indicate non-radiative processes.

Materials and Methods

2.4.1.1 Substrate binding assays

Substrate binding assays were performed with 1.5 μM purified CaiT protein by recording the tryptophan fluorescence emission. Concentrations of potential substrates such as glycine betaine (Sigma), γ-butyrobetaine (Lonza Group Ltd.), L-carnitine (Sigma), or choline (Sigma) were used at concentrations ranging from 0.25 to 40 mM. The fluorescence emission was recorded using the F-4500 Fluorescence Spectrophotometer (Hitachi). The excitation wavelength was set to 295 nm and emission spectra between 305 nm and 390 nm were recorded. The slit width was set to 2.5 nm or 5 nm for measurements of excitation or emission, respectively. Each fluorescence emission spectrum represents the mean of 3 to 5 recordings. The mean value and standard deviation at maximum emission at 342 nm was plotted for each substrate concentration. Data were fitted to the Michaelis-Menten equation as previously described (Jung et al., 2002) and apparent K_M values were calculated using the program Origin 7.5. To determine the sodium dependency of substrate binding experiments were carried out at different NaCl concentrations (1 – 50 mM) or without added NaCl.

2.4.2 Freeze-fracture electron microscopy

A small sample of proteoliposome solution (1.5 – 2 μl, final lipid concentration 60 mg/ml) was pipetted between two copper plates serving as a sample holder. The sample was rapidly frozen in liquid ethane at -180 °C. Freeze-fracture of proteoliposomes was carried out with a BAF 060 machine (Bal-Tec) at a temperature of -130 °C and a pressure of 2×10^{-7} mbar. The resulting fracture planes were shadowed with platinum / carbon followed by pure carbon shadowing to reinforce the replica. The replica was floated with sulfuric acid (40 - 50 % v/v) for 10 hours, washed several times with water for cleaning and analyzed in an EM208S electron microscope (FEI Company). Images were collected on a 1K × 1K slow-scan CCD camera (TVIPS). All freeze-fracturing and imaging steps were performed by F. Joos (MPI of Biophysics).

2.4.3 X-ray crystallography

X-ray crystallography is the prevalent method of determining the three-dimensional (3D) structure of biological macromolecules at atomic resolution. It is based on the fact that macromolecules diffract X-rays in a pattern, which is dependent on their orientation in 3D space. In crystals, atoms or molecules are arranged in regular repeating arrays, which act as diffraction grating from which X-rays are scattered in particular directions. This scattered radiation can be detected as a two-dimensional array of spots of particular positions and intensities forming the diffraction pattern. The extent of the diffraction pattern is directly correlated with the degree of internal crystal order. The more structurally uniform and precisely arraged the molecules are in the crystal, the higher is the resolution to which the pattern extends (McPherson, 2004).

2.4.3.1 Prediction of 3D crystallization feasibility

The success-rate of forming highly ordered, well diffracting crystals is dependent on the internal protein properties, which are defined by the amino acid sequence. The *XtalPred* server (http://ffas.burnham.org/XtalPred, (Slabinski *et al.*, 2007b)) estimates, based on the amino acid sequence, the success rate for protein production and crystallization by analyzing several predicted protein features such as sequence length, isoelectric point, hydrophobicity index or gravy index, which is the ration of the sum of the hydrophobicity values for each residue in the protein to the actual protein length (Canaves et al., 2004), length of the longest disordered fragment, instability index, predicted percentage of coil secondary structure, number of residues in the predicted coil-coiled regions, predicted transmembrane helices, percentage of insertions in multiple alignments, predicted long low-complexity regions, predicted signal peptide and net protein charge (Slabinski *et al.*, 2007a) . One has to keep in mind that the individual protein features are compared to proteins that are available in databases like the TargetDB, (Chen *et al.*, 2004)), the Protein

Materials and Methods

Structure Initiative (PSI, (Norvell and Machalek, 2000)) and the protein data bank (PDB). However, there are still only comparatively few membrane protein structures (252 unique membrane protein X-ray structures, PDB, June 2010) deposit in the PDB compared to the number of soluble protein structures (66 072 soluble protein X-ray structures, PDB, June 2010), which makes an unbiased estimation for membrane proteins difficult. Nevertheless, the analysis of the results from the crystallization feasibility predictions between homologs helps to find the one homolog that might be more likely to from well-ordered, high diffracting crystal for structure determination.

2.4.3.2 Crystallization of CaiT

Protein crystals can grow from a supersaturated solution (Figure 2-3) after nucleation has taken place. A stable nucleus is an ordered aggregate of protein molecules that provides a surface at which free molecules can bind *via* hydrogen bonds, electrostatic or hydrophobic interactions. The crystal grows by incorporating free protein molecules into its lattice (McPherson, 2004).

Materials and Methods

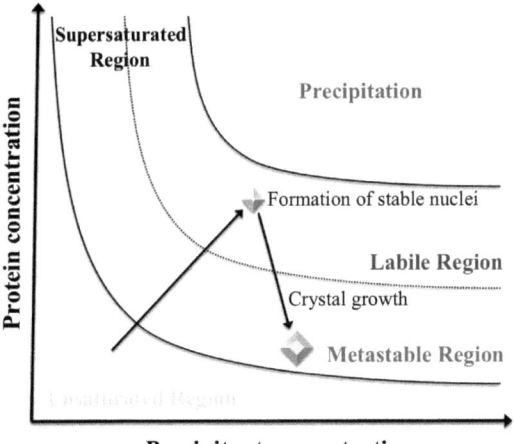

Figure 2-3 | Phase diagram for protein crystal growth

The phase diagram shows three regions. In the unsaturated region (yellow) crystals cannot form because a solid phase would immediately dissolve. The supersaturated region (black) is divided into two subregions. In the metastable region (green) crystal growth is possible, but protein crystal nucleation occurs spontaneously at higher saturation levels in the labile region (blue). A continuous increase in protein or precipitant concentration beyond the labile region leads to protein precipitation (red).

The phase diagram (Figure 2-3) shows that reaching supersaturation depends on the protein as well as on the precipitant concentration. Additional parameters which influences protein crystal nucleation and growth are: the type of precipitant e.g. salts (NaCl or $(NH_4)_2SO_4$), organic compounds, e.g. polyethylene glycol (PEG) or jeffamine, or non-volatile organic solvents, e.g. 2-methyl-2,4-pentanediol (MPD) or 2,5-hexandiol, the pH of the solution, the temperature and the crystallization method itself.

CaiT was crystallized using the hanging drop or sitting drop vapour diffusion methods (Figure 2-4). The protein solution is mixed with reservoir solution and placed above the reservoir in a closed chamber. Since the concentration of precipitant in the reservoir is initially higher than in the drop, water molecules leave by vapour diffusion until the osmolarity of the drop and the reservoir becomes equal. The loss of

Materials and Methods

water in the drop leads to a continuous increase of protein and precipitant concentration.

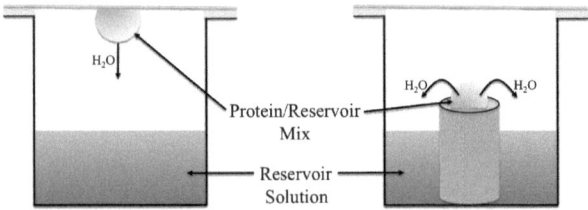

Figure 2-4 | Schematic representation of the hanging drop and sitting drop vapour diffusion method

Since the concentration of precipitant in the reservoir solution is initially higher than in the protein drop mixed with reservoir solution, water molecules leave the drop by vapour diffusion until the osmolarity of the drop and the reservoir are equal.

To find initial crystallization conditions for CaiT, commercial crystallization screens were used, notably the Index™, Crystal Screen™ and Crystal Screen II™ screens from Hampton, the MbClass™ and MbClass II™ screens from Qiagen, the MemStart™ and MemSyst™ screens from Molecular Dimension and the JBScreen Basic™ and JBScreen Membrane™ screens from Jena Bioscience. All screening crystallization trials were performed in 96-well sitting drop plates using the Mosquito™ (TTPLabtech, UK) pipetting robot. The reservoirs of the 96-well sitting drop plate were manually filled with 70 µl of crystallization solution. Typically, the crystallization drops were prepared using 500 nl of protein solution (5 mg/ml) plus 250 – 500 nl of reservoir solution. The crystallization plates were incubated at different temperatures (4°C, 6°C, 8°C, 10°C, 12°C, 16°C, 18°C, 22°C, 25°C).

Crystallization conditions were optimized in 24-well plates (Hampton Research) using the hanging drop vapour diffusion method. For crystallization 500 µl of optimized crystallization conditions (Table 2-21) was used as reservoir solution. The crystallization drops were set up with 2 µl of CaiT protein solution (2.5 – 7.5 mg/ml) mixed with 1 µl reservoir solution.

Materials and Methods

Table 2-21 | **Optimized crystallization conditions**

Component	Concentration
MgAc$_2$ or CaAc$_2$, pH 4.4 – 5.5	50 mM
PEG 400	15 – 26 % (w/v)
NaCl	0 – 275 mM

2.4.3.3 Seeding

Macro- and microseeding experiments were performed to optimize CaiT crystals. For a macroseeding experiment, small crystals (100 – 150 µm) grown in one condition were washed in stabilizing condition, which usually contained a higher PEG 400 concentration. Then the crystals were transferred into a non-stabilizing solution so that the crystals began to dissolve. The roundish looking crystals were then transferred to new crystallization drops, which had been pre-equilibrated for several hours (usually 2 – 12 hours).

For a microseeding experiment one big crystal (300 – 400 µm) was first washed in stabilizing solution and then transferred into a test tube containing 50 µl of stabilizing solution. The seed stock solution was vortexed and centrifuged (4 °C, 2000 g, 10 sec). The seed stock was serially diluted (1:5, 1:10, 1:20 and 1:50) and 0.5 µl of the diluted seed stocks were added to pre-equilibrated crystallization drops. The crystallization drops contained either different protein or precipitant concentrations or were pre-equilibrated for different time spans (usually 2 – 12 hours).

2.4.3.4 Detergent screens

Detergent Screening Kits[TM] 1 – 3 (Hampton Research) were used to find new or additional detergents for optimizing CaiT crystal growth. Each Detergent Screen[TM] contains 24 unique detergents at 10× of the critical micelle concentration (CMC). For detergent screening 5 µl of a double concentrated optimized crystallization condition

Materials and Methods

was mixed with 1.5 µl of the detergent solution. Water was added to a final volume of 10 µl (detergent crystallization solution). Crystallization drops were set up using 2 µl of CaiT protein solution (2.5 – 7.5 mg/ml) mixed with 1 µl detergent crystallization solution. The reservoir solution contained 500 µl of the optimized crystallization condition without additional detergent.

Conditions that produced crystals were reproduced several times (usually six times) using the same detergent crystallization solution. X-ray diffraction of the crystals was tested and compared in order to decide which detergents indeed optimized crystal growth.

2.4.3.5 Additive screens

Additives are small molecules that can affect crystallization by changing protein-protein or protein-solvent interactions. The Additive Screen^{TM} (Hampton Research) contains 96 different compounds at 10× the recommended drop concentration. Additives were screened using the pipetting robot *Mosquito*^{TM} (TTPLabtech, UK). The reservoirs of the 96-well sitting drop plate were manually filled with 70 µl of crystallization solution. Usually, 175 nl of reservoir solution was mixed with 75 nl of additive, which was then added to 500 nl of protein solution. Conditions that yielded protein crystals were reproduced in 24-well plates. X-ray diffraction of the crystals was tested and compared to find the best additives.

2.4.3.6 Heavy atom screens

To overcome the phase problem (Section 2.4.7.1) anomalous scatterers are used in the multiple or single isomorphous replacement (MIR/SIR) method and the multiple or single anomalous dispersion (MAD/SAD) method. Both of these approaches require at least one heavy atom derivative of the protein. The search for a

Materials and Methods

heavy atom that binds without damaging the protein or the crystal is as empirical as determining the initial crystallization conditions.

For heavy atom screening, the Heavy Atom Screen Pt^{TM}, Heavy Atom Screen Au^{TM}, Heavy Atom Screen Hg^{TM}, Heavy Atom Screen M1^{TM} and Heavy Atom Screen M2^{TM} screen from Hampton and the Tantalum Cluster Derivatization Kit^{TM} from Jena Bioscience were used. All heavy atom screening trials were performed in 24-well plates (Hampton Research) by hanging drop vapour diffusion at 4 °C.

2.4.3.7 Cryocrystallography

Cryocrystallography (Haas, 1968) is based on the collection of X-ray diffraction at liquid nitrogen temperature (100 K). The main advantage of the method is the greately reduced rate of radiation damage to the protein crystal during data collection at cryogenic temperatures. Since the diffusion of radicals created in the crystals by X-ray radiation is limited in the amorphous phase of flash-cooled crystals, their lifetime in the beam is increased by about 100-fold (Nave and Garman, 2005; Weik and Colletier, 2010).

To protect the protein crystal from damage caused by ordered ice formation during flash freezing, the crystal is transferred into a solution of cryoprotectant agents before freezing. Cryoprotectant agents, such as PEG 400 or glycerol, penetrate into the solvent-filled channels of protein crystals (Garman and Doublié, 2003; Garman and Schneider, 1997) and interact extensively with water molecules. These agents thereby prevent the formation of crystalline ice and promote the formation of vitreous ice during flash freezing. In contrast to crystalline ice, vitreous ice does not expand on formation and therefore does not damage the protein crystals (McPherson, 1999).

CaiT crystals were cryoprotected with MgAc$_2$ or CaAc$_2$ buffered PEG 400 (30 – 35 % (w/v)). To reduce the osmotic shock, the total volume of 2 μl cryosolution was added stepwise (incubation time 1 – 2 min) to the crystallization drop. After an

Materials and Methods

additional incubation time of 5 – 10 minutes the crystals were fished with cryo-loops (Hampton Research; Molecular Dimension) and flash frozen in liquid nitrogen.

2.4.4 Principles of X-ray crystallography

A crystal is built up from regularly repeating units. The smallest unit of a crystal is the asymmetric unit (AU). It is the minimum group of atoms whose positions, by applying crystallographic symmetry operations, generate the complete content of the unit cell. The AU contains the unit cell origin and the primary generating symmetry element(s). The unit cell is the smallest repeating volume that when periodically translated forms the crystal lattice. The unit cell is defined by three lattice constants, a, b, c, and by three angles α, β, γ.

Translation and rotation are symmetry operations, which define the lattice symmetry of a crystal. A unit cell and as a consequence also the crystal lattice can only have 2-fold, 3-fold, 4-fold and 6-fold rotational symmetry. The multiplicity n of the rotation axis and the rotation angle ϕ are related by $\phi = (360/n)°$. In a trigonal unit cell (P3), a rotation around the 3-fold symmetry axis (▲) involves three consecutive rotations by 120°. It is not possible to fill a plane with e.g. regular pentagons, and no true crystal lattice with 5-fold symmetry exists.

By applying all allowed symmetry operations a crystal can be regarded as a 3D space lattice, an infinite array of points where its neighbours surround each point in an identical way.

Protein X-ray crystallography usually uses X-rays with wavelengths in the range of 0.6 – 2.5 Å. The electromagnetic radiation can interact with electrons and atoms in a crystal and diffraction occurs when the dimensions of the diffracting objects (e.g. bond lengths of molecules) are comparable to the wavelength of the radiation. The diffraction pattern of a crystal is a list of angles at which reflections are observed. Bragg's law (Figure 2-5) explains the quantitative relation between the interplanar lattice spacing of a crystal (d_{hkl}) and the scattering angles (θ) at which constructive interference (reflection) occurs from a specific lattice point. The

Materials and Methods

reciprocal space construction with the Ewald sphere represents the 3D illustration of Bragg's law. The Ewald sphere of radius $1/\lambda$ illustrates which reflections will be imaged by the detector and explains why the 2-dimensional diffraction pattern shows diffraction spots in separate ellipses (Figure 2-5; (Dauter, 1999)).

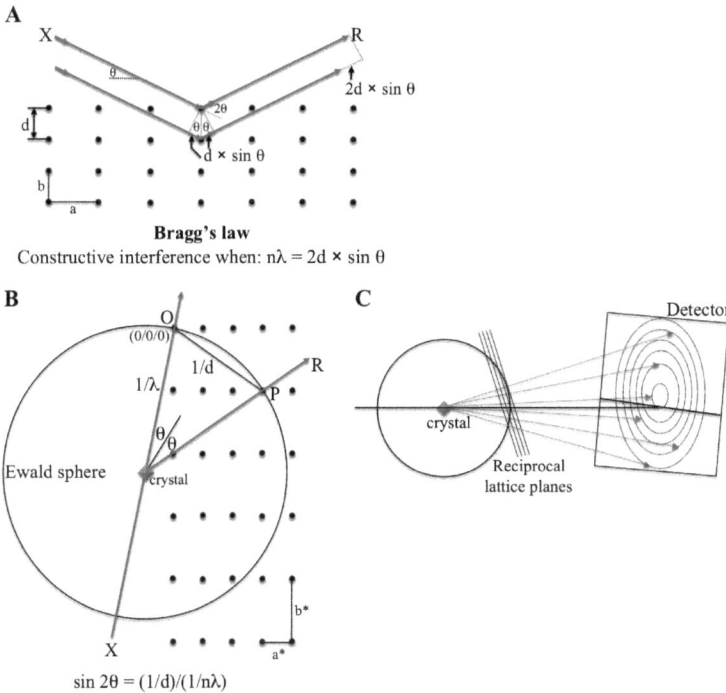

Figure 2-5 | Conditions for constructive interference

Bragg's law (**A**) describes at which conditions diffraction is observed. Constructive interference, which produces strong diffraction, occurs when the difference in path length of reflected rays (R) from successive crystal planes is equal to an integral number (n) of wavelengths of the incident X-rays (X). Strongly reflected rays emerge from planes of spacing d only at angles θ (Bragg angles) for which $2d \times \sin\theta = n\lambda$.

The reciprocal space construction with the Ewald sphere (**B**) shows that reflections are observed on the detector when a reciprocal lattice point (P) coincides with the Ewald sphere of radius $1/\lambda$. As the crystal rotates in the X-ray beam (X), various reciprocal lattice points come in contact with the Ewald sphere each producing a reflecting beam (R). This produces a characteristic diffraction pattern with diffraction spots in separated ellipses (**C**). This figure is adapted from (Dauter, 1999).

Materials and Methods

2.4.5 Data collection

Diffraction of CaiT crystals was tested in-house at the MicroMax^{TM}-007HF diffractometer with R-AXIS IV^{++} detector (RIGAKU Americas) or at the FR-E$^+$ SuperBright diffractometer with Saturn 944$^+$ detector (RIGAKU Americas). The first device operates at 40 kV and 30 mA, whereas the latter one operates at 45 kV and 55 mA.

High-resolution data sets of native and derivative crystals were collected either the Max-Planck beamline of the Swiss Light Source (SLS, Villigen, Switzerland) or at the *European Synchrotron Radiation Facility* (ESRF, Grenoble, France).
Before collecting a complete data set, three test images were collected each 60° or 90° apart. These three images were indexed and data collection strategy was optimized. High-resolution data were usually collected in two passes of different resolution. First a complete low-resolution data set was collected before high-resolution data were recorded.

2.4.6 Data processing

Indexing and data integration was performed using MOSFLM (Leslie, 1992) or XDS (Kabsch, 1993). Data were initially processed in $P1$ using XDS. The correct space group was determined using POINTLESS (Evans, 2006) and data were subsequently scaled and merged using SCALA (CCP4, 1994). Beside the completeness of the data set, the signal-to-noise ratio (I/σI) as well as the R_{merge} factor (Equation 2-2) were used to judge data quality.

Materials and Methods

Equation 2-2:

$$R_{merge} = \frac{\sum_{hkl} \sum_i \|F_{hkl}\| - |F_{hkl}(i)\|}{\sum_{hkl} \sum_i |F_{hkl}(i)|}$$

|F$_{hkl}$ (i)| is the value of structure factor amplitude of the i-th reflection with Miller indices hkl. |F$_{hkl}$| is the final value of the structure factor amplitude

2.4.7 Phasing

2.4.7.1 The phase problem

In order to calculate the electron density of a protein three parameters are refined for each diffraction spot: its index (hkl), intensity (I$_{hkl}$) and its phase angles (α_{hkl}) of the reflections. From a diffraction pattern only two of these parameters are measured.

The complex structure factor from the contents of the unit cell is a function of the electron distribution with the unit cell volume V.

Equation 2-3:

$$\mathbf{F}_{hkl} = V \cdot \int_x \int_y \int_z \rho_{xyz} \exp[2\pi i(hx + hy + lz)] dx dy dz$$

$\mathbf{F}_{hkl}$ is the Fourier integral of ρ_{xyz}. Since ρ_{xyz} is the inverse Fourier transform of $\mathbf{F}_{hkl}$ and the reciprocal space of a crystal is discrete, the integral can be replaced with a discrete triple summation. The electron density can then be calculated as follows:

Materials and Methods

Equation 2-4:

$$\rho_{xyz} = \frac{1}{V} \cdot \sum_x \sum_y \sum_z \mathbf{F}_{hkl} \exp[-2\pi i(hx + ky + lz)]$$

Since $\mathbf{F}_{hkl} = |F_{hkl}| \exp[i\alpha_{hkl}]$, Equation 2-4 can be written as:

Equation 2-5:

$$\rho_{xyz} = \frac{1}{V} \cdot \sum_x \sum_y \sum_z |F_{hkl}| \exp[-2\pi i(hx + ky + lz) + (i\alpha_{hkl})]$$

The structure factor amplitude $|F_{hkl}|$ can be calculated from the intensities I of the reflection spots: $I_{hkl} \propto |F_{hkl}|^2$. In contrast, the phase angle α_{hkl} cannot be derived from the diffraction pattern directly. Several methods have been developed to solve this so-called phase problem in X-ray crystallography.

2.4.7.2 Methods to solve the phase problem

1. Multiple Anomalous Dispersion (MAD) or Single Anomalous Dispersion (SAD)

The diffraction intensities of a native protein structure are normally centrosymmetric. That means two reflections hkl and -h-k-l are related to one another by inversion through the origin (O) and the measured intensities, as well as the calculated structure factors, of the two reflections are equal (Figure 2-6). This relationship is called Friedel's law.

Materials and Methods

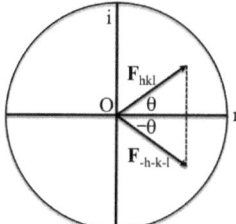

Figure 2-6 | Vector diagram for a Friedel pair

For centrosymmetric structures, two reflections have equal structure factors with equal amplitudes $|F_{hkl}|$ = $|F_{-h-k-l}|$ and opposite phases θ_{hkl} = $-\theta_{hkl}$. (r = real axis, i = imaginary axis)

The introduction of atoms that absorb X-rays of a particular wavelength into a native protein structure can break Friedel's law. When the wavelength of X-rays is chosen to be close to an absorption edge of the atom that was introduced into the native protein structure, some of the photons are first absorbed by the electrons of the this atom before they are re-emitted. This leads to a phase shift of the scattered photons ($\theta_{hkl} \neq -\theta_{hkl}$) and the amplitudes of the reflections hkl and -h-k-l are not equal anymore ($|F_{hkl}| \neq |F_{-h-k-l}|$) (Figure 2-7). This inequality of symmetry-related reflections is called anomalous dispersion (Blow and Rossmann, 1961; Hendrickson and Lattman, 1970). The symmetry-related reflections (hkl and -h-k-l) are now called Bijvoet pair.

Materials and Methods

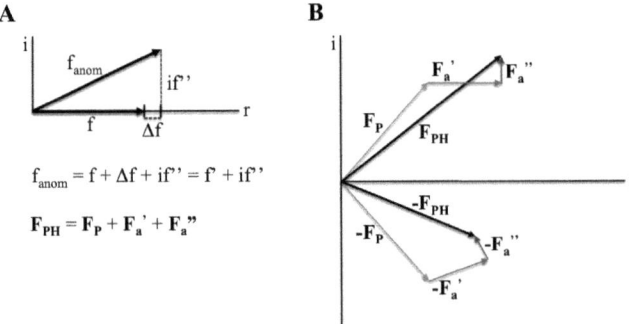

Figure 2-7 | Vector diagram for anomalous scattering conditions

The atomic scattering factor for an anomalous scatterer (f_{anom}) consists of two parts: a real part f' and the imaginary part if" (**A**). Introducing an anomalous scatterer into a native protein structure causes the break down of Friedel's law (**B**). F_P and $-F_P$ are the structure factors for the Friedel's pair, whereas F_{PH} and $-F_{PH}$ are the structure factors for the Bijvoet pair. F_a' and F_a" are the real and imaginary parts of structure factors of the anomalous scatterer, respectively (Blow and Rossmann, 1961; Hendrickson and Lattman, 1970).

Collection of MAD data at different wavelengths can provide very accurate experimental phases. The variation in scattering properties of heavy atoms as a function of incident X-ray wavelength result both in differences between reflection intensities at different wavelengths (dispersive differences) as well as differences between the intensities of the Bijvoet pairs (anomalous differences) at the same wavelength (González, A., 2003). In order to optimize both anomalous differences and dispersive differences in a MAD experiment, data collection at three wavelengths is necessary. These three wavelengths are the peak wavelength (λ_p), the wavelength at the inflection point (λ_i) and the remote wavelength (λ_r) (Figure 2-8).

Data collection at the peak wavelength, where the value of the imaginary component of the anomalous scattering factor if" reaches a local maximum, optimizes the anomalous differences.

The dispersive differences are proportional to the difference in the real component $\Delta f'$ ($\Delta f' = f_{rem}' - f_{min}'$). The dispersive differences are ideal when f_{min}' reaches the minimum value at the inflection point and the value for f_{rem}' is high at a wavelength far away from the edge (remote wavelength).

Materials and Methods

In order to optimize the phasing power of a three-wavelength MAD experiment, the anomalous scattering factor at the remote wavelength (f_{rem}') should be such that it maximizes the quantity if' × Δf' (González, 2003). In a two-wavelength MAD experiment, the best strategy is to optimize the dispersive (Δf'), rather than the anomalous (if') differences. This means data collection at the inflection and remote wavelength is preferred (González, 2003; González et al., 1999). For a SAD experiment it is best to optimize the anomalous differences and to collect data at the peak wavelength (Dauter et al., 2002).

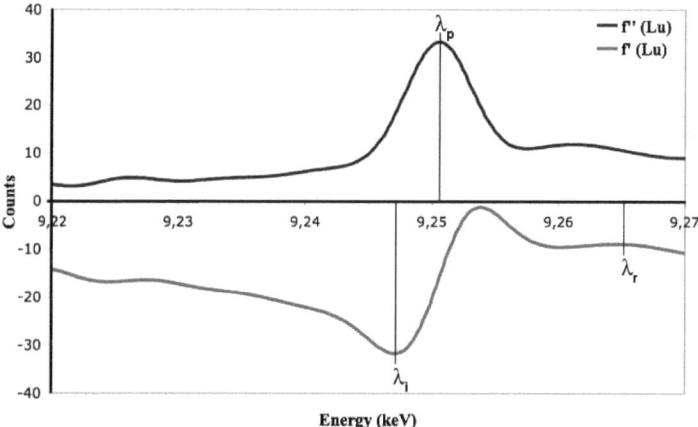

Figure 2-8 | Fluorescence scan of a Lu-containing CaiT crystal
At peak wavelength (λ_p), f'' reaches a local maximum, the anomalous differences of the Bijvoet pairs are maximal. At the inflection point (λ_i) the value for f' is minimal. The dispersive differences Δf' (Δf' = f_{rem}' − f_{min}') become optimal when the value for f_{rem}' at the remote wavelength (λ_r) is high.

The heavy atom positions can be determined from the Harker vectors in the Harker sections of the difference Patterson map. After calculation of the structure factor amplitudes for F_H, F_{PH} and F_P ($F_{PH} = F_P + F_H$), the phase angles of the heavy atoms (α_H) and of the protein (α_P) can be determined and subsequently an initial electron density map of the protein structure can be calculated.

Materials and Methods

For determination of initial heavy atom positions the programs SHELXD (Schneider and Sheldrick, 2002), SHARP (de La Fortelle and Bricogne, 1997; Vonrhein et al., 2007) and SOLVE (Terwilliger and Berendzen, 1999) were used in this work.

2. Molecular Replacement (MR)

The MR method (Rossmann and Blow, 1962) can be used if a crystal structure of a protein with a homologous amino acid sequence is known. The phases of the structure faktors of the known protein (search model) can be used to estimate the phases of the unknown protein. In the Fourier transform (Equation 2-4, Equation 2-5), the amplitudes are obtained from the native intensities of the unknown protein ($|F_{hkl}$ (unknown protein)$|$) and the phases are taken from the search model ($i\alpha_{hkl}$ (search model)). During refinement the initial R values (R/R_{free}) are usually high, which represent the differences between the phases of the search model and the phases of the unknown protein. During the refinement process, the phases should change from those of the search model to those of the unknown protein and the R values should decrease.

For the MR method, the Patterson function of the search model is used to explore the Patterson function of the crystal structure to be determined. The orientation and the precise position of the unknown protein in the unit cell are determined by rotation and translation (Figure 2-9). In the rotational search, the search model is placed within a $P1$ unit cell whose dimensions guarantee that the volume contains only intramolecular vectors (also called self vectors). The self vectors only depend on the orientation of the molecule and can therefore be used to determine the spatial orientation of the known and unknown protein and the orientation with respect to each other. Once the orientations of the molecules have been determined, the translation vector needed to superimpose the correctly oriented molecules is calculated (Drenth, 2007; Messerschmidt, 2007).

Materials and Methods

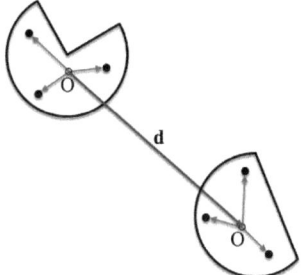

Figure 2-9 | Schematic description of the MR method
The Patterson function of two molecules separated by a spatial movement can be formulated as X_2 = $[C]X_1 + d$, where X_1 is the position of molecule 1 and X_2 is the positions of molecule 2, $[C]$ is the rotation matrix and d is the translation vector (blue arrow). The orientations of the molecules are determined by their self vectors (red arrows around the origin O). The rotation matrix $[C]$ of the molecules is characterized by three angles α, β, γ (Euler angles). By varying the rotation around these angles, the molecules can be brought into any desired angular orientation. The translation vector defines the spatial movement that superimposes the two molecules (Blow, 2005; Messerschmidt, 2007).

In this work MR was performed to determine the structure of PmCaiT with a poly-alanine model of BetP (PDB accession code 2WIT). The EcCaiT structure was determined afterwards by using PmCaiT as search model for MR. In both cases the program PHASER (McCoy *et al.*, 2007) was used. Model bias was minimized by using the prime-and-switch program (Terwilliger, 2004). The primary density was improved by density modification in RESOLVE (Terwilliger, 2000).

2.4.8 Phase improvement, model building and refinement

After an initial set of protein phases is obtained, the next step is to interpret the map by automatically placing 'dummy' atoms and refining their positions. These 'dummy' atoms are used to define bond lengths and the number of neighbouring atoms of the initial model. A further phase improvement is achieved by density modification with solvent flattening or solvent flipping combined with histogram matching. If more than one molecule is found in the asymmetric unit (Section 2.4.4) and these molecules are related by one or more noncrystallographic symmetry (NCS) operators, NCS averaging can be used to improve the protein phase angles by averaging the electron density about the different related subunits.

Model building and model refinement is an repetitive alternating process of real space fitting the model into the electron density and global reciprocal space restrained refinement of the model's positional parameters. After every cycle the refined phase angles can be used to calculate a new electron density map. The $2F_{obs} - F_{calc}$ Fourier coefficient amplitudes map displays the atomic model with normal weights and indicates errors in the model by its contribution of the difference Fourier map (Messerschmidt, 2007). Iterative cycles of model building and/or model refinement improve the electron density map in such a way that amino acid side chains can be correctly assigned. The parameters of the initial model (x, y, z coordinates, overall and/or individual atomic displacement parameter (B-factor), scaling factor) are adjusted to minimize the difference between the calculated and observed structure factors. The quality of the crystallographic model can be estimated from the crystallographic R-factor (Equation 2-6; (Drenth, 2007)).

Equation 2-6:

$$R_{work} = \frac{\sum_{hkl} ||F_{obs}| - k|F_{calc}||}{\sum_{hkl} |F_{obs}|}$$

During refinement geometric and stereochemical restrains (bond length, bond angles and dihedral angles) are applied to the structural model. The conformation of the main chain folding is verified by the Ramachandran plot (Ramachandran et al., 1963) in which the dihedral angles ϕ and ψ are plotted against each other for each residue. The angles are correct when they lie in the energetically favoured regions of secondary structures such as α-helices, β-sheets and defined turn structures (Messerschmidt, 2007).

Materials and Methods

Both the EcCaiT and PmCaiT structures were refined with iterative cycles of manual rebuilding in COOT (Emsley and Cowtan, 2004) and O (Jones *et al.*, 1991), and simulated annealing followed by maximum likelihood-based energy minimization and isotropic B-factor refinement in PHENIX (Adams *et al.*, 2002). The validation of both final models, PmCaiT and EcCaiT, was performed using the program PROCHECK (Laskowski *et al.*, 1993).

2.4.9 Figures

Diagrams of the CaiT structure were generated using the programs POV*Script+* (Fenn *et al.*, 2002) and Povray (http:/www.povray.org), Bobscript (Esnouf, 1999) as well as PyMol (DeLano, 2002). Surface potentials were made using PyMol. Superpositions were carried out with the SSM (Krissinel and Henrick, 2004) superposition routine of COOT (Emsley and Cowtan, 2004).

3 Results

In this work, three different homologs of the carnitine transporter CaiT were investigated. *Proteus mirabilis* CaiT (PmCaiT) and *Escherichia coli* CaiT (EcCaiT) share a sequence identity of 87 %, whereas the sequence identity of PmCaiT and *Salmonella typhimurium* CaiT (StCaiT) is 84 % (Figure 6-2). The twelve transmembrane helices (TM1 – TM12) of the CaiT homologs share a sequence identity of 95 %. Non-conserved residues are confined mostly to the loop regions connecting the transmembrane helices in the protomer, or to the N- and C-terminal ends (Figure 6-2).

The three homologs of CaiT were chosen to enhance the probability of high-resolution 3D structure determination. This was succesfull with the *P. mirabilis* homolog. The 3D structure of CaiT was then used as a basis for functional analysis with the wildtype protein and mutations of functional key residues. The combination of structural and functional analysis revealed details of the substrate transport mechanism for CaiT.

Unless explicitly mentioned and described, the CaiT homologs were treated equally. Results that are representative for all three CaiT homologs are shown only once for one of the three homologs.

Results

3.1 Expression and purification

3.1.1 Expression of CaiT in *E. coli*

3.1.1.1 Native CaiT protein

The *caiT* gene was cloned into the pET-15b vector using the restriction enzymes *Nde*I and *Xho*I. Restriction enzyme digests followed by sequencing determined the success of cloning. Plasmids containing the correctly inserted *caiT* gene were transformed into *E. coli* BL21(DE3) pLysS cells for expression. Primary expression tests in 100 ml LB media showed an expression level (Figure 3-1) sufficient for crystallization and activity studies.

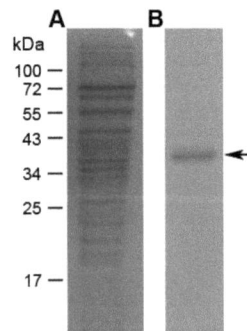

Figure 3-1 | SDS-PAGE gel and Western blot of PmCaiT test expression
The cells of a 100 ml culture were harvested by centrifugation. After diluting the cell pellet to an OD_{600} of 1, the cells were disrupted and an aliquot of 10 µl was loaded onto the SDS gel (**A**). Western blot analysis (**B**) using an α-His-tag antibody resulted in a single protein band (arrow), which runs between 43 kDa and 34 kDa. It was previously shown for EcCaiT (Vinothkumar, 2005) that CaiT migrates in the SDS gel at this size.

To optimize the expression of native CaiT, several conditions were tested including growth media, expression strains, growth temperature, IPTG concentration for induction and time of induction. The best protein expression was obtained with *E. coli* BL21(DE3) pLysS cell in 2×YT media. Cells were grown at 37 °C to an OD_{600} of 0.6 to 0.9 before protein over-expression was induced with 0.5 mM ITPG. After induction, the growth temperature was reduced to 30 °C and the cells were further cultured for 4 hours.

3.1.1.2 SeMet-labelled CaiT protein

To produce SeMet-labelled PmCaiT and EcCaiT, *E. coli* BL21(DE3) RIL-X cells were transformed with the pET-15b vector containing Pm*caiT* or Ec*caiT*.

The SeMet-labeled protein production in flasks was problematic since the cells did not grow to a sufficient cell mass. Methionine-auxotrophic cells cultured in flasks with *SelenoMet* medium, selection marker and SeMet only grew to an OD_{600} of 0.6 – 0.9. After induction, the cell mass either stagnated or decreased which led to a low protein yield even though the expression of SeMet-labeled CaiT was not the problem. The fed-batch fermentation strategy was more efficient for uncoupling the accumulation of cell mass from the induction and expression of SeMet-labeled protein (Studts and Fox, 1999).

The production of SeMet-CaiT was optimized using the fed-batch fermentation (Section 2.3.2.2). The best results were obtained when the *E. coli* BL21(DE3) RIL-X cells were grown in *SelenoMet* Medium at 37 °C to an OD_{600} of 1.3 to 1.7. Protein over-expression was induced with 0.5 mM IPTG and cells were further grown for 18 – 20 hours at a temperature of 30 °C. During the fermentation period the cell density of the 3 l culture increased to an OD_{600} of 2.4 – 2.6.

Results

3.1.2 Purification of CaiT

3.1.2.1 Native CaiT protein

The purification of CaiT involved three steps: solubilization of the protein (Section 2.3.4.2), immobilized metal affinity chromatography (IMAC, Section 2.3.4.3) and preparative size exclusion chromatography (SEC, Section 2.3.4.4). Although the procedure of protein purification (Section 2.3.4) was identical for all constructs, two strategies were followed depending on what the protein was subsequently used for. When the protein was used for crystallization, it was solubilized and purified in Cymal-5 and HEPES buffer in presence of NaCl. In contrast, protein used for binding and transport measurements was solubilized and purified in DDM, sodium-free Tris buffer and without any added NaCl as it was one aim to test the sodium dependence of substrate binding and transport.

The best crystallization results were obtained when the protein was solubilized in 2 % (w/v) Cymal-5, 25 mM HEPES pH 7.5, 100 mM NaCl, 10 % (v/v) glycerol and 1 mM TCEP (Figure 3-2). After solubilization, 50 mM imidazole pH 8.0 was added to the solubilized membrane fraction before it was applied to the Ni^{2+} column. The imidazole helped to avoid unspecific protein binding to the column. The column was washed with two different buffers. The first washing buffer, which contained 50 mM imidazole pH 8.0 and 10 % (w/v) glycerol, removed impurities from the sample (Figure 3-2). The second washing step removed the glycerol, which was used during solubilization and in the IMAC binding and the first washing buffer. The protein was eluted from the Ni^{2+} column with 200 mM imidazole pH 8.0 (Figure 3-2) and loaded directly onto a PD-10 column to remove the imidazole from the protein which tended to induce protein aggregation. The buffer used for eluting the protein from the PD-10 column contained 0.12 % (w/v) Cymal-5, 25 mM HEPES pH 7.5, 100 mM NaCl and 1 mM TCEP. The most concentrated elution fractions (Figure 3-2) usually contained 5 mg/ml CaiT, which was adequate for crystallization. Fractions with less protein were pooled and concentrated to 5 mg/ml. The two protein samples with a concentration of 5 mg/ml were not pooled but handled separately to check whether or not the additional concentration of the protein had an effect on the crystallization behaviour.

For activity measurements the protein was solubilized in 2 % (w/v) DDM, 50 mM sodium-free Tris pH 7.5, 10 % (v/v) glycerol and 1 mM TCEP for 1 – 2 hours at 4 °C. The purification was performed as described above but with buffers containing 0.05 % DDM, 50 mM sodium-free Tris pH 7.5 and 1 mM TCEP was used. The imidazole and glycerol concentrations used for the Ni^{2+} column were kept as before. For activity studies the protein had either to be reconstituted into liposomes or could be used directly in detergent solubilized form. In both cases the protein concentration did not need to be as high as for crystallization. To avoid protein aggregation due to the low ionic strength of the buffer, the protein concentration was kept at 2 – 3 mg/ml.

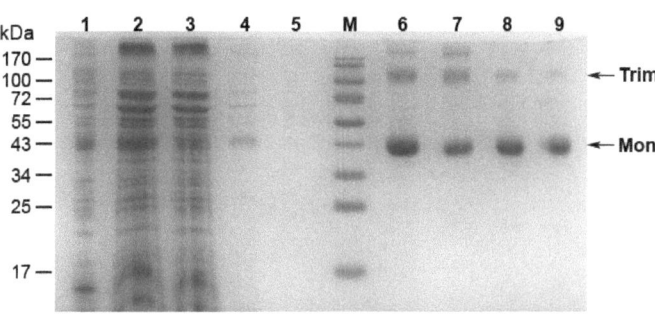

Figure 3-2 | SDS-PAGE gel of a PmCaiT purification
E. coli BL21(DE3) pLysS cells were grown in 2×YT medium. Protein over-expression was induced with 0.5 mM IPTG. After expression cells were harvested, broken and membranes (**1**) were isolated. The solubilized membrane fraction (**2**) was loaded onto a Ni^{2+}-Sepharose column. The flow-through (**3**) was discarded and the column was washed twice (**4, 5**) before the protein was eluted (**6, 7**) with 200 mM imidazole pH 8.0 in fractions of 500 – 1000 µl. To avoid protein aggregation caused by the high imidazole concentration, the protein was directly loaded onto a size exclusion column and eluted (**8**) in fractions of 200 – 500 µl. The purified protein (**9**) was either used for crystallization or activity studies. (**M**) PageRuler™ Prestained Protein Ladder (Fermentas). Previous studies showed that CaiT froms trimers in detergent solution (Vinothkumar, 2005; Vinothkumar *et al.*, 2006). In the Coomassie stained SDS-PAGE gel both the monomeric form (arrow, Mon) and the trimeric form (arrow, Trim) of CaiT is detectable.

Results

The purification of EcCaiT included an additional step in which the His_6-tag was cleaved off by incubating the protein 1 – 2 hours at 4 °C with the immobilized serine protease thrombin. The removal of the His_6-tag from EcCaiT is detectable by a shift of the protein band on the SDS-PAGE gel (Figure 3-3).

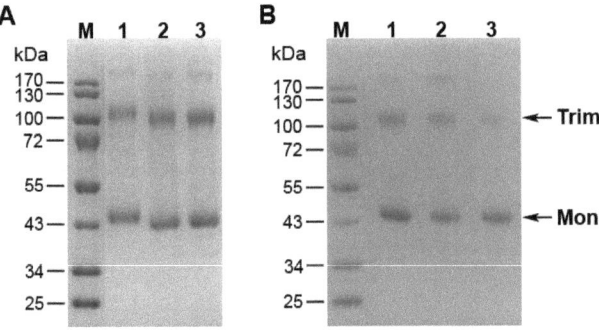

Figure 3-3 | SDS-PAGE gel of purified EcCaiT and StCaiT

The SDS-PAGE gels show the final steps in the purification of EcCaiT (**A**) and StCaiT (**B**). The protein was eluted from the Ni^{2+}-Sepharose column (**1**) and desalted by size exclusion chromatography (**2**). The purified protein (**3**) was used for crystallization or activity measurements. EcCaiT was additionally incubated with thrombin after the protein was eluted from the Ni^{2+} column. The cleavage of the His_6-tag is detectable in the Coomassie stained SDS-PAGE gel by a shift of the protein band. (**M**) PageRuler™ Prestained Protein Ladder (Fermentas).

3.1.2.2 SeMet-labelled CaiT

SeMet-labelled protein was used only for crystallization. To optimize the crystallization procedure, several parameters were tested including different detergents, detergent concentrations, reducing agents, concentrations of reducing agents and different buffers. Although the SeMet-labelled protein was considered as a new protein and crystallization conditions were newly screened, the best crystallization results were obtained using a purification protocol that was similar to that used for native protein.

Results

SeMet-labelled CaiT was solubilized in 1.5 % (w/v) Cymal-5, 25 mM HEPES pH 7.5, 100 mM NaCl, 10 % (v/v) glycerol and 2 mM TCEP. After solubilization, 50 mM imidazole pH 8.0 was added to the solubilized membrane fraction before it was applied to a Ni^{2+} column. As for the native protein the Ni^{2+} column was washed with two buffers containing 50 mM imidazole pH 8.0, first with 10 % (w/v) glycerol and then without. The protein was eluted from the Ni^{2+} column with 200 mM imidazole pH 8.0 and loaded directly onto a PD-10 column (Figure 3-4). The buffer used for eluting the protein from the PD-10 column contained 0.12 % (w/v) Cymal-5, 25 mM HEPES pH 7.5, 100 mM NaCl and 2 mM TCEP (Figure 3-4). Since the protein concentration was very low, all protein-containing elution fractions had to be pooled and concentrated. The protein was concentrated to 5 mg/ml (Figure 3-4), which was the ideal concentration for crystallization.

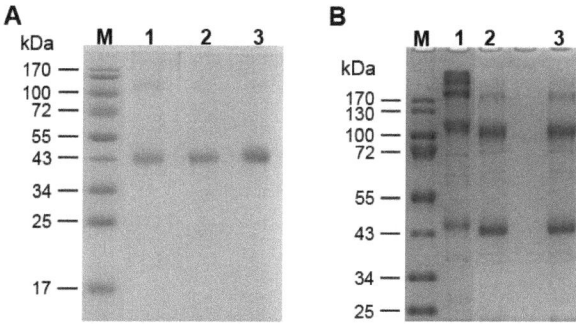

Figure 3-4 | SDS-PAGE gel of purified SeMet-PmCaiT and SeMet-EcCaiT
The SDS-PAGE gels show the final steps in the purification of SeMet-PmCaiT (**A**) and SeMet-EcCaiT (**B**). The solubilized membrane fraction containing SeMet-CaiT was loaded onto a Ni^{2+} column. After washing, the protein was eluted (**1**) with 200 mM imidazole pH 8.0 and directly loaded onto a size exclusion column. The eluted protein (**2**) was concentrated to 5 mg/ml. The purified and concentrated protein (**3**) was used for crystallization. (**M**) PageRuler™ Prestained Protein Ladder (Fermentas).

3.1.3 Sample quality of CaiT

After purification, the protein sample was tested for stability and monodispersity by BN-PAGE (Section 2.3.7.1) and analytical size exclusion chromatography (SEC) (Section 2.3.4.4). A stable and monodisperse protein sample is essential for biochemical measurements such as binding and transport activity and for growing highly ordered, well diffracting crystals for structure determination. The two different methods that were used are appropriate for membrane proteins. The BN-PAGE uses Coomassie G250, which induces a charge shift on the protein and stays tightly bound to it, and ε-aminoaproic acid, which improves the solubility of membrane proteins during electrophoresis. The analytical SEC was pre-equilibrated in detergent to ensure that the protein remains in solution during the run.

3.1.3.1 Blue-Native PAGE

4 – 12 % gradient native gels were loaded with 3 – 7.5 µg of purified protein. The PAGE was started at 80 V to concentrate the samples within the stacking gel. Afterwards the PAGE was run 4 – 5 h at a constant voltage of 150 V. Native EcCaiT and StCaiT usually separated into two bands, PmCaiT into three bands during electrophoresis (Figure 3-5 **A**). The first protein band migrated to a size between 140 kDa and 232 kDa, the second protein band ran to a size between 232 kDa and 440 kDa, while the third band of PmCaiT ran at about 440 kDa according to the native protein marker (HMW Native Marker; GE Healthcare). The first protein band indicates the CaiT trimers (169.8 kDa) plus additional bound detergent molecules. The second detectable band contains protein that has the appropriate size for CaiT hexamers (339.6 kDa) plus bound detergent molecules. This band probably represents CaiT trimer-trimer–interactions that are caused by the high protein concentration within the gel. The third band of the PmCaiT sample in the BN gradient gel could be a CaiT nonamer.

On the BN-PAGE of SeMet-labelled protein (Figure 3-5 **B**, **C**) only one protein band was clearly detectable. This band migrated to a protein size between 140 kDa and 232 kDa, indicating CaiT trimers. Compared to the native CaiT the SeMet-labelled protein shifts closer to the 140 kDa band. A reason for the different migration behaviour might be the polarizability of the selenium atoms in the protein.

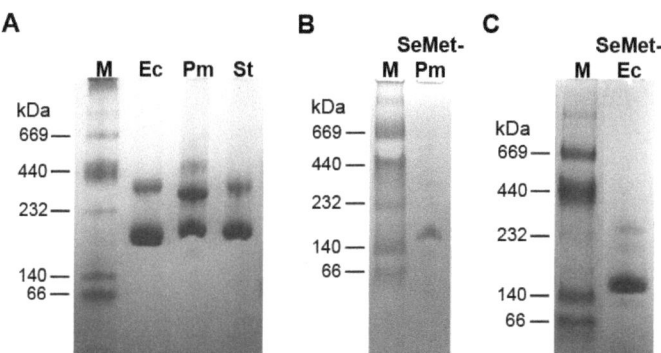

Figure 3-5 | BN-PAGE gradient gel of purified native and SeMet-labelled CaiT

After purification each sample was checked for stability and monodispersity by BN-PAGE. The gel was loaded with 3 – 7.5 µg of purified protein. The BN-PAGE gradient gel (4 – 12 %) of native CaiT (**A**) shows two prominent bands for each EcCaiT (**Ec**), PmCaiT (**Pm**) or StCaiT (**St**) sample. The first band in the gradient gel indicates a protein size between 140 kDa and 232 kDa, while the second protein band runs between 232 kDa and 440 kDa according to the high molecular weight (HMW) Native Marker (**M**; GE Healthcare). In the PmCaiT sample a third band is detectable that migrates at about 440 kDa. The protein band between 140 kDa and 232 kDa is also prominent in the gels loaded with SeMet-derivatives of PmCaiT (**B**, **SeMet-Pm**) or EcCaiT (**C**, **SeMet-Ec**).

The protein band that migrates between 232 kDa and 140 kDa has the appropriate size for CaiT trimers surrounded by a detergent micelle. The higher protein band between 232 kDa and 440 kDa is likely to be two interacting trimers, as the size is appropriate for a hexameric CaiT plus surrounding detergent molecules. The third band in the PmCaiT sample that runs at about 440 kDa might represent a CaiT nonamer. The interactions between two or more CaiT trimers is likely an effect of the high protein concentration in the gel.

3.1.3.2 Analytical size exclusion chromatography (SEC)

Analytical SEC was the second method to check the stability and dispersity of the protein sample. For analytical SEC, a Superose^{TM} 6 (3.2/30) column was used, which had a bed volume of 2.4 ml and an estimated void volume of 0.72 ml (approximately 30 % of the total bed volume, GE Healthcare). After equilibration with the appropriate buffer (Table 2-13), the purified protein sample (50 – 100 µg) was loaded onto the column. The separation of the sample was performed with a constant flow rate of 50 µl/min and fractions of 100 µl were collected using the Ettan LC system.

CaiT eluted as a single peak with a retention volume of 1.58 – 1.62 ml (Figure 3-6). Occasionally protein was also detected in the void volume, which was considered to be aggregated. Samples, which showed a peak in the void volume, were subjected to ultra-centrifugation (4 °C, 100,000 g, 20 min) to pellet the aggregates before the sample was reanalyzed. Samples that eluted in a single peak were used for activity measurements or for 3D crystallization.

Results

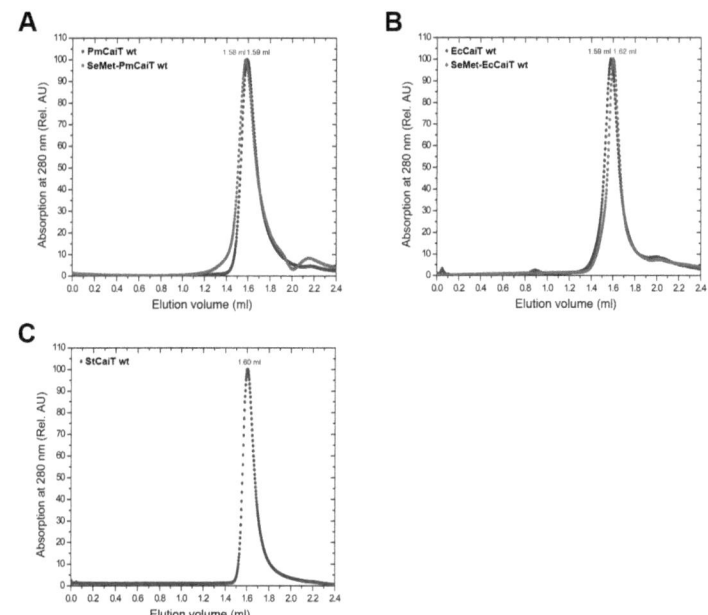

Figure 3-6 | Analytical size exclusion chromatography of CaiT

Analytical SEC was performed using a Superose™ 6 3.2/10 column, which had a total bed volume of 2.4 ml and an estimated void volume of 0.72 ml. Purified protein samples (0.5 – 1 µg/µl) were loaded onto the pre-equilibrated column. Separation was performed using a constant flow rate of 50 µl/min. Native protein (blue line) usually eluted in a single peak at an elution volume of 1.58 ml – 1.60 ml. SeMet-labelled protein (red line) eluted with the same retention volume as native protein but contained an additional small shoulder beside the main peak. The shoulder was caused by the higher concentration of the reducing agent TCEP used in the SeMet-CaiT buffer as the native EcCaiT (**B**) also showed this peak when it was purified in the same buffer. Analytical SEC elution profile of native PmCaiT wildtype (PmCaiT wt, blue line) and SeMet-labelled PmCaiT wildtype (SeMet-PmCaiT wt, red line) (**A**). Native EcCaiT wildtype (EcCaiT wt, blue line) and SeMet-labelled EcCaiT wildtype (SeMet-EcCaiT wt, red line) (**B**). Native StCaiT wildtype (StCaiT wt, blue line) (**C**).

Both methods, BN-PAGE and analytical SEC, demonstrate that both native and SeMet-labelled CaiT protein is stable and monodisperse in detergent solution after purification. This makes CaiT suitable for activity and structural studies.

Results

3.2 3D crystallization

The expression of CaiT in BL21(DE3) pLysS cell was optimized in this work to produce sufficient amounts of protein for 3D crystallization. The highly expressed protein was folded and did not aggregate. After purification a protein yield of 2 – 3 mg/(l culture) was obtained. The high expression level, the stability and the monodispersity of the protein were a prerequisite for high-resolution 3D structural studies.

3.2.1 3D crystallization feasibility prediction of CaiT

Structural studies were undertaken on all three CaiT homologs of which EcCaiT was the first. The EcCaiT project was taken over from K. R. Vinothkumar, who already could show that EcCaiT forms 2D crystals in the lipid membrane environment which tended to stack (Vinothkumar, 2005; Vinothkumar *et al.*, 2006). Since EcCaiT formed 2D crystals, it was promising to analyze its ability to from 3D crystals. K. R. Vinothkumar also grew initial 3D crystals of EcCaiT (Vinothkumar, 2005).

The *XtalPred* server (Section 2.4.3.1) calculates an individual protein production and crystallization feasibility score and places the result into one of five success rate categories: optimal, suboptimal, average, difficult and very difficult. The prediction score for the 3D crystallization feasibility of EcCaiT (Figure 3-7) was placed into the "very difficult" bin. The server found 388 homologs which all had the same score. However, the *XtalPred* predictions for EcCaiT, PmCaiT (Figure 3-7, Figure 3-8) and StCaiT (Figure 3-9) showed differences that might influence their crystallization behaviour. The biggest difference between the CaiT homologs was in the hydrophobicity index or gravy index (blue arrow) and in the isoelectric point (pI, black arrow). The gravy index for both EcCaiT and StCaiT was 0.72 (Figure 3-7, Figure 3-9) that for PmCaiT it was 0.68 (Figure 3-8). It has been reported that a gravy

Results

index of 0.1 is the optimum for protein crystallization (Slabinski et al., 2007a), indicating that PmCaiT, with the lowest gravy index, should crystallize better than the other two homologs. The pI for EcCaiT is 8.54, for PmCaiT 8.41 and for StCaiT 8.67. The crystallization success rate was shown to be higher for neutral or slightly acidic proteins (Slabinski et al., 2007a). All CaiT homologs are basic proteins but the pI of PmCaiT (Figure 3-8) seemed slightly more favourable compared to the pI of EcCaiT and StCaiT (Figure 3-7, Figure 3-9).

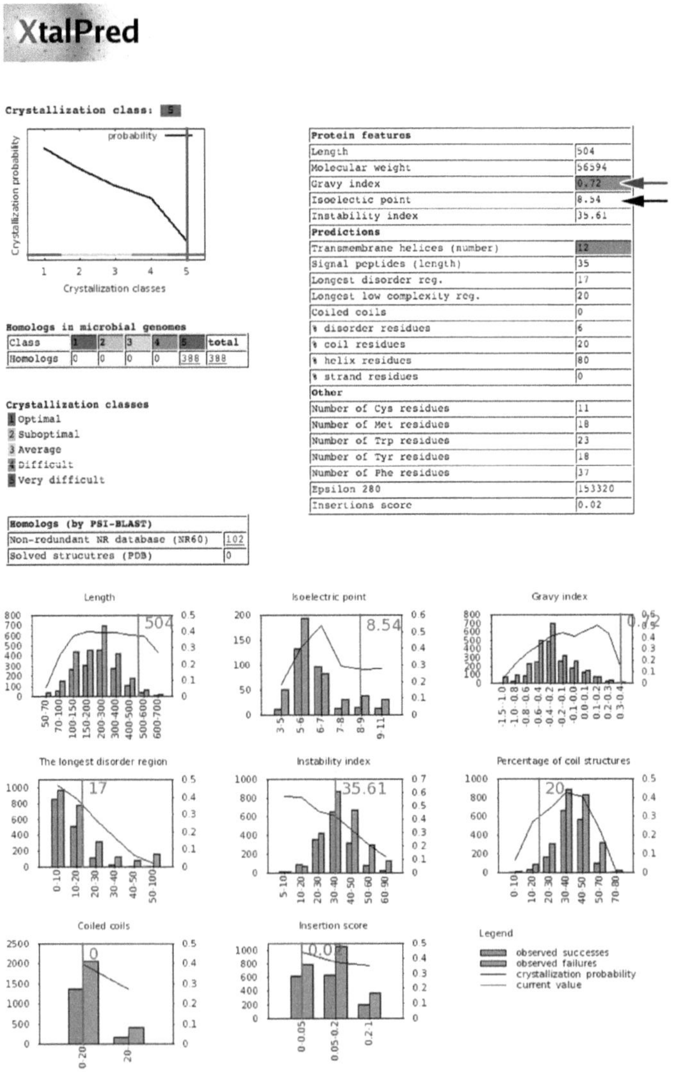

Results

Figure 3-7 | Prediction of EcCaiT crystallization probability

The crystallization probability was estimated from the EcCaiT protein sequence using the *XtalPred* server (http://ffas.burnham.org/XtalPred).

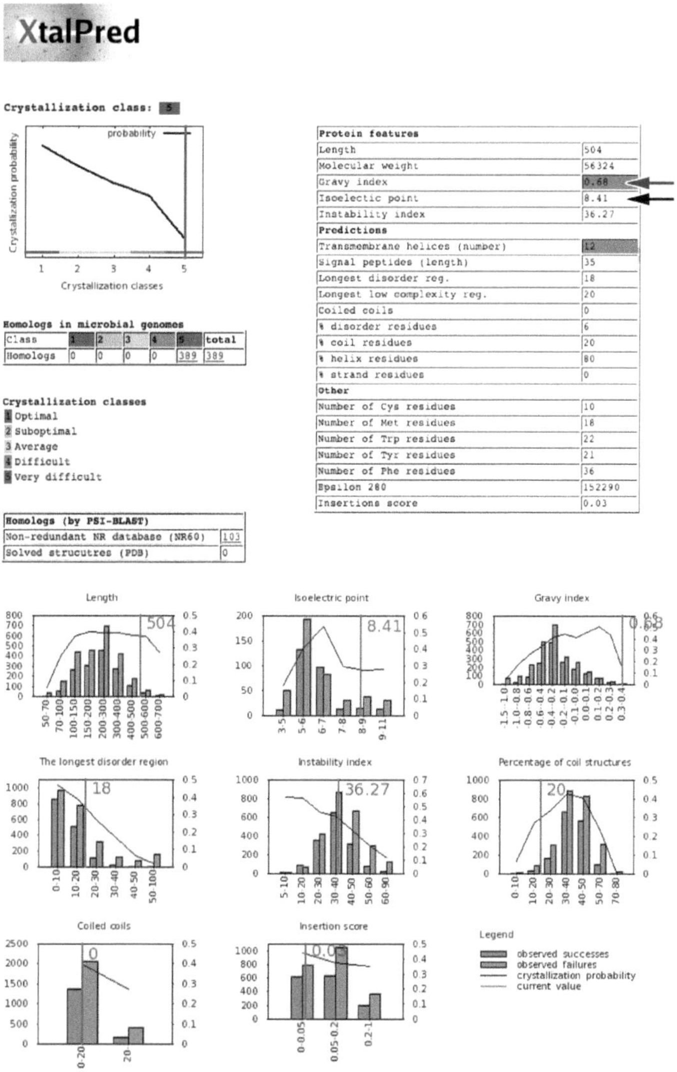

Figure 3-8 | Prediction of PmCaiT crystallization probability

The crystallization probability was estimated from the PmCaiT protein sequence using the *XtalPred* server (http://ffas.burnham.org/XtalPred).

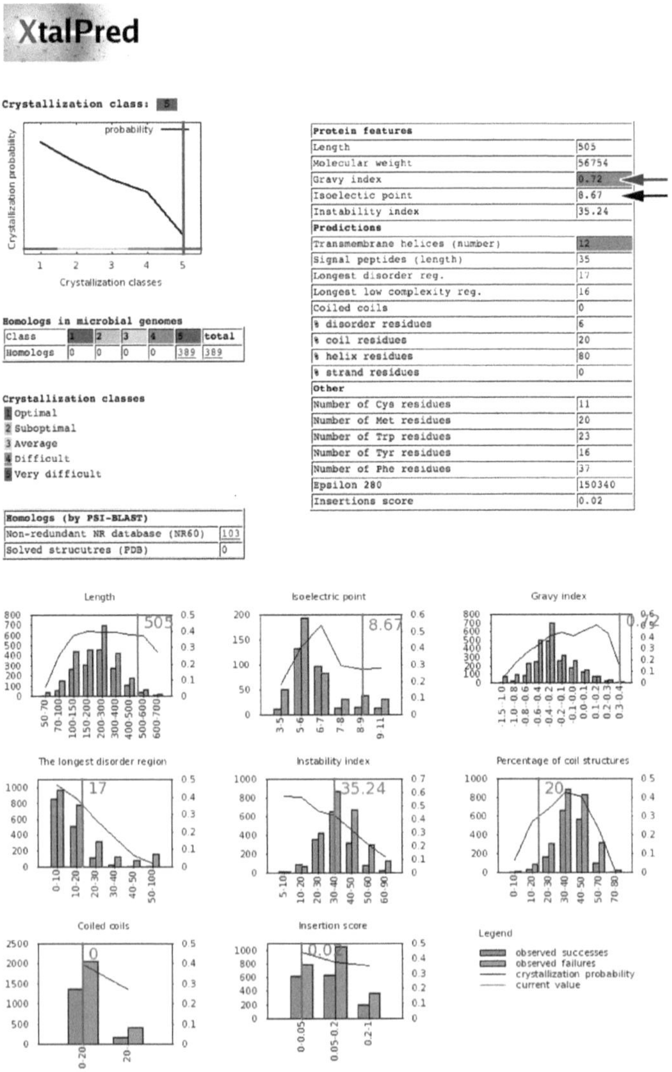

Results

Figure 3-9 | Prediction of StCaiT crystallization probability

The crystallization probability was estimated from the StCaiT protein sequence using the *XtalPred* server (http://ffas.burnham.org/XtalPred).

3.2.2 3D crystallization trials

Initial crystallization trials were performed using commercially available screens from Hampton, Qiagen and Jena Bioscience. All three CaiT homologs yielded initial crystals in several different conditions. The best initial crystals were obtained with the condition 0.1 M sodium acetate trihydrate pH 4.5 and 25 % (w/v) PEG3350 from the Index^{TM} screen (Hampton). However, the crystal quality was quite different. Initial EcCaiT crystals (Figure 3-10 **A**) diffracted to a resolution of ~ 9 Å. The diffraction limit for initial PmCaiT crystals (Figure 3-10 **B**) was ~ 5 Å, while initial StCaiT crystals (Figure 3-10 **C**) did not show any diffraction spots (MacroMax^{TM}-007HF diffractometer, R-AXIS IV^{++} detector, in-house). The sharp-edged crystals of EcCaiT and PmCaiT showed a rhomboidal to rectangular morphology. StCaiT crystals had no well-defined edges but formed long triangles.

Protein samples that were concentrated to 5 mg/ml resulted in many but small crystals whereas samples that eluted from the PD-10 column with a protein concentration of 5 mg/ml resulted in few but large crystals.

Results

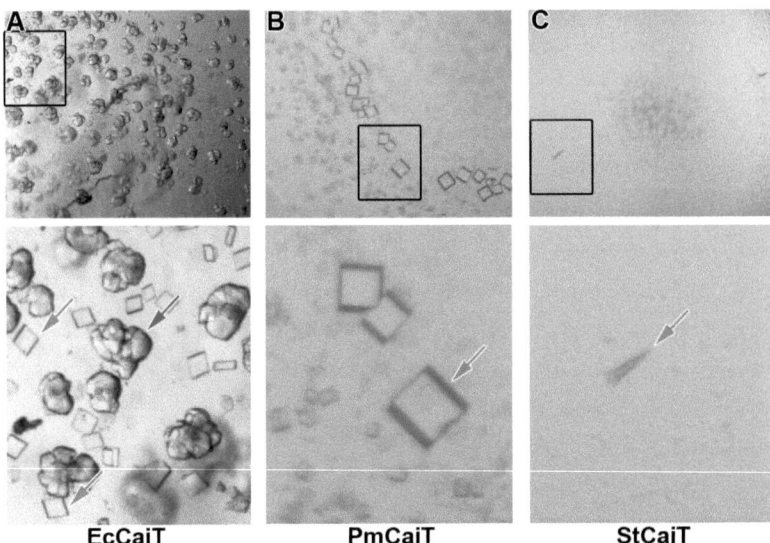

Figure 3-10 | Initial crystallization trials of EcCaiT, PmCaiT and StCaiT
Initial crystals (green arrow) of EcCaiT (**A**), PmCaiT (**B**) and StCaiT (**C**) were grown in 100 mM NaAc pH 4.5 and 25 % (w/v) PEG3350. The initial EcCaiT protein buffer contained DDM at a concentration of 0.05 % (w/v), which was too high and resulted in the formation of spherulits (red arrow).

The crystals observed were tested to establish that they were indeed of CaiT. It was checked if the dissolved protein crystal had the same migration behaviour in the SDS-PAGE gel and in the BN-PAGE gel as freshly purified CaiT in solution. This was necessary to exclude the possibility that a protein impurity was crystallized and to find out if the CaiT protein remained stable during crystallization.

Three initial crystals (50 – 100 nm in the longest direction) were fished in a nylon loop and transferred into a drop of stabilizing solution, which had the same composition as the protein-free crystallization solution with an additional 5 % (w/v) PEG400. The crystals were washed twice before they were transferred into 5 µl of protein buffer that contained detergent but no PEG400. After the crystals had been dissolved completely in the protein buffer, 2 µl of the solution were analyzed by

SDS-PAGE (Figure 3-11 **A**), while the other 3 µl were analyzed by BN-PAGE (Figure 3-11 **B**).

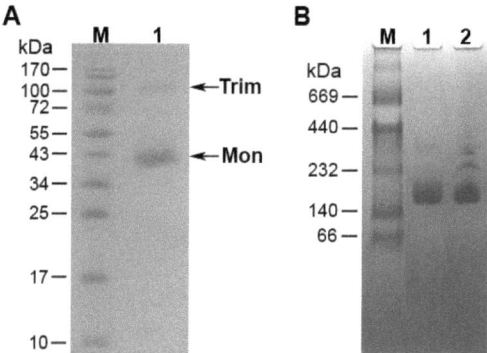

Figure 3-11 | SDS-PAGE gel and BN-PAGE gradient gel of dissolved CaiT crystals

Three of the initial crystals were fished, washed and transferred into a 5 µl drop of protein buffer. It was checked with the light microscope that the crystals had dissolved completely. 2 µl of the solution were used for SDS-PAGE, the other 3 µl were used for BN-PAGE. The dissolved protein crystal solution (**1**) migrated in the SDS-PAGE gel (**A**) as two bands. Both protein bands have the characteristic size for the monomeric (**Mon**) and the trimeric (**Trim**) form of CaiT. (**M**) is the PageRuler™ Prestained Protein Ladder (Fermentas). The dissolved protein crystal solution (**1**) and freshly purified CaiT (**2**) migrated to the same position in the BN-PAGE gradient gel (**B**), but only one band was visible in the sample from the CaiT crystals that migrated to a size between 140 kDa and 232 kDa, which is the appropriate size for CaiT trimers. The size of the CaiT bands was determined in relation to the HMW Native Marker (**M**; GE Healthcare).

SDS-PAGE and BN-PAGE confirmed that the initial crystals contained CaiT protein and demonstrated that CaiT was stable in the crystallization solution.

The initial crystallization conditions were optimized by testing different buffers and varying the pH of the solution, trying different but related salt types in various concentrations, altering the concentration and chain length of the precipitant PEG, screening different detergents and detergent concentrations and carrying out the crystallization in the presence or absence of different concentrations of additives.

To improve the crystal quality, different crystallization methods, like sitting and hanging drop vapour diffusion or the under-oil batch method were tested. Additional parameters like the protein concentration (2.5 – 7.5 mg/ml), the protein-to-reservoir ratio (1:0.5, 1:1, 1:2, 0.5:1 and 2:1), the reservoir volume (250 – 1000 µl) and the crystallization temperature (25 °C, 22 °C, 18 °C, 16 °C, 12 °C, 10 °C, 8 °C, 6 °C and 4 °C) were varied. Crystal seeding (Section 2.4.3.3) was attempted. Various parameters were modified in order to optimize seeding. The parameters were: varying the seed stock concentration, protein concentration, precipitant concentration and incubation period.

The best crystals grew at 4 °C by the hanging drop vapour diffusion method. The crystallization drops were set up with a protein concentration of 5 mg/ml and a protein-to-reservoir ratio of 2:1. The total reservoir volume was 500 µl and crystals grew in the mother liquor containing: 50 mM MgAc$_2$ or CaAc$_2$ pH 4.5 – 5.5, 0 – 200 mM NaCl, 19 – 30 % (w/v) PEG400, which were set up in two 96-well fine screens. The conditions that yielded the best crystals for the respective CaiT homolog were set up in 24 well plates.

PmCaiT crystallized at 4 °C in both MgAc$_2$ (Figure 3-12 **A**) and in CaAc$_2$ (Figure 3-12 **B**) at pH 5.0 and grew to a final size of 200 – 500 µm in the longest dimension within 3 – 10 days. The change of crystal morphology caused by the two ions Mg^{2+} and Ca^{2+} is striking. PmCaiT crystals that were grown in MgAc$_2$ showed a diffraction limit of 2.8 Å, whereas PmCaiT crystals that were grown in CaAc$_2$ diffracted to 2.3 Å. Both crystal forms diffracted isotropically. Although native PmCaiT produced well-ordered crystals, it was not possible to generate crystals that diffracted similarly well of SeMet-labelled PmCaiT.

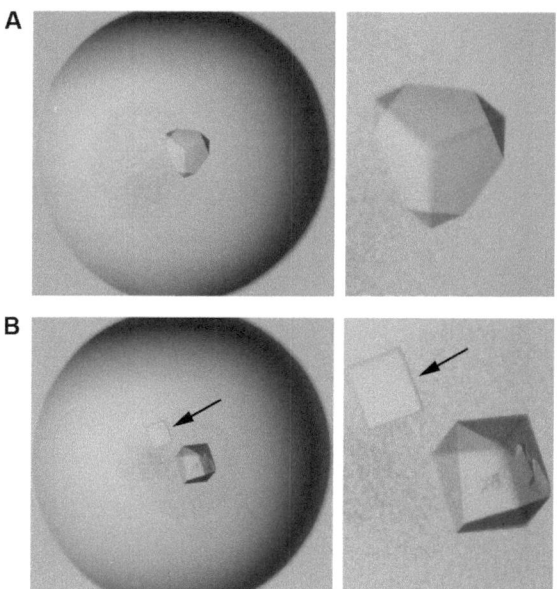

Figure 3-12 | Optimized PmCaiT crystals

Optimized crystallization conditions for PmCaiT yielded crystals with different morphologies and diffraction properties. Triangular pyramidal crystals (**A**) were grown at 4 °C in 50 mM MgAc$_2$ pH 5.0, 50 – 100 mM NaCl and 21 – 23 % (w/v) PEG400. These crystals diffracted to a resolution of 2.8 Å. Rectangular PmCaiT crystals (**B**) were grown at 4 °C in 50 mM CaAc$_2$ pH 5.0, 50 – 100 mM NaCl and 21 – 23 % (w/v) PEG400 and diffracted to a resolution of 2.3 Å. One of these highly-diffracting crystals (black arrow) was used to solve the CaiT structure.

EcCaiT crystallized at 4 °C in MgAc$_2$ (Figure 3-13 **A**) and in CaAc$_2$ (Figure 3-13 **B**) at pH 4.8 – 5.2. The crystals grew within 1 – 2 weeks to their final size of 300 – 600 µm in the longest dimension. In contrast to PmCaiT, the two ions, Mg^{2+} and Ca^{2+}, did not have the same striking effect on the crystal morphology for EcCaiT as for PmCaiT. In both conditions the crystals had a triangular shape. However, the Ca^{2+} containing conditions produced more regular shaped crystals. EcCaiT crystals that were grown with Mg^{2+} diffracted isotropically to a resolution of 3.8 – 4 Å, while crystals grown with Ca^{2+} diffracted isotropically to a resolution of 3.5 – 3.8 Å.

The rhomboidal EcCaiT crystals that grew in the initial screen condition were reproducible but reached a maximum crystal size of only ~100 µm in the longest direction and showed a diffraction limit of ~5.5 Å.

SeMet-labelled EcCaiT crystals (Figure 3-13 **C**) grew only with Ca^{2+} at pH 4.8 – 5.0. The crystals grew within 1 – 2 weeks to a size of 100 – 200 µm, had sharp edges and a triangular shape. These crystals diffracted to 5.5 – 6.0 Å.

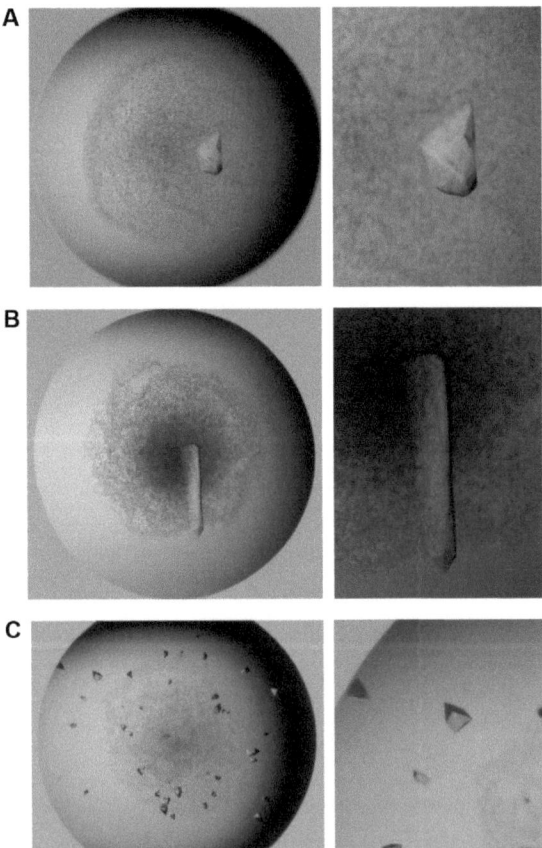

Figure 3-13 | Optimized EcCaiT crystals

Optimized crystallization conditions for EcCaiT produced crystals with different triangular morphologies. EcCaiT crystals with less well-defined triangular shape (**A**) were grown at 4 °C in 50 mM MgAc$_2$ pH 4.8 – 5.2, 100 mM NaCl and 23 – 25 % (w/v) PEG400. These crystals diffracted to a resolution of 3.8 – 4.0 Å. Long triangular pyramidal EcCaiT crystals (**B**) were grown at 4 °C in 50 mM CaAc$_2$ pH 4.8 – 5.0, 100 mM NaCl and 23 – 25 % (w/v) PEG400 and diffracted to a resolution of 3.5 – 3.8 Å. Triangular SeMet-EcCaiT crystals (**C**) were grown at 4 °C in 50 mM CaAc$_2$ pH 4.8 – 5.0, 100 mM NaCl and 25 – 28 % (w/v) PEG400. These crystals showed a diffraction limit of 5.5 – 6.0 Å.

StCaiT crystallized at 4 °C only in MgAc$_2$ at pH 4.5 – 4.8 (Figure 3-14). StCaiT crystals grew within 1 – 2 weeks to their final size of 300 – 400 μm in the longest dimension. These crystals had a long triangular morphology and although they appeared robust, they fractured when touched with the Micro-Brush™ (Hampton). The best crystals diffracted anisotropically to a resolution of 3.9 – 4.5 Å.

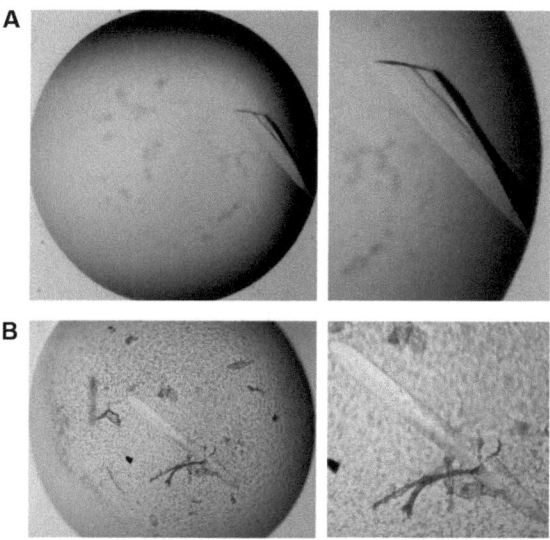

Figure 3-14 | Optimized StCaiT crystals
Optimized crystallization conditions for StCaiT produced crystals with a long triangular morphology. StCaiT crystals were grown at 4 °C in 50 mM MgAc$_2$ pH 4.5 – 4.8, 100 mM NaCl and 21 – 23 % (w/v) PEG400 (**A**) or in 50 mM MgAc$_2$ pH 4.5 – 4.8, 200 mM NaCl and 25 – 28 % (w/v) PEG400 (**B**). The best StCaiT crystals diffracted anisotropically to a resolution of 3.9 – 4.5 Å.

For data collection, crystals were flash-frozen directly in liquid nitrogen. Before freezing, the crystals were cryo-protected by a stepwise increase of the PEG400 concentration in the crystallization drop. The cryo-solution added to the crystallization drop was identical to the crystallization condition protein-free solution from which the crystals grew but contained 40 % (w/v) PEG400.

3.3 Structure determination

3.3.1 Data collection and data processing

Diffraction data were collected on the beamlines X10SA at the *Swiss Light Source* (SLS, Villigen, Switzerland) and ID 14-eh3 at the *European Synchrotron Radiation Facility* (ESRF, Grenoble, France).

Diffraction quality was judged from 3 images that were collected 45 ° or 60 ° apart. Crystals showing isotropic diffraction to an appropriate resolution were chosen for data collection. The starting angle and optimal collection strategy was determined using the program STRATEGY, which is part of the image processing program MOSFLM (Leslie, 1992).

The diffraction pattern of a PmCaiT crystal is shown in Figure 3-15. The crystal diffracted isotropically to a resolution of 2.3 Å and the data set was used for the structure determination of PmCaiT.

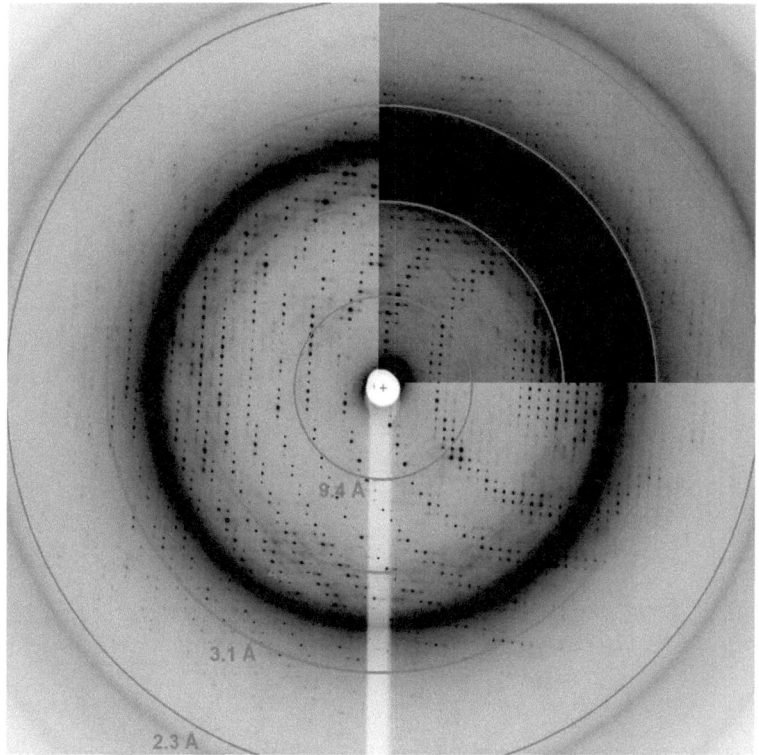

Figure 3-15 | Diffraction pattern of a PmCaiT crystal

The data set was collected on the beamline X10SA at the SLS (Villigen, Switzerland) from the smaller PmCaiT crystal shown in Figure 3-12 **B**. Data were collected to a resolution of 2.3 Å. The upper right quarter of the diffraction pattern was adjusted to a higher contrast level to show the high-resolution diffraction spots.

The diffraction pattern of an EcCaiT crystal is shown in Figure 3-16. This crystal diffracted isotropically to a resolution of 3.5 Å and this data set was used for the EcCaiT structure determination.

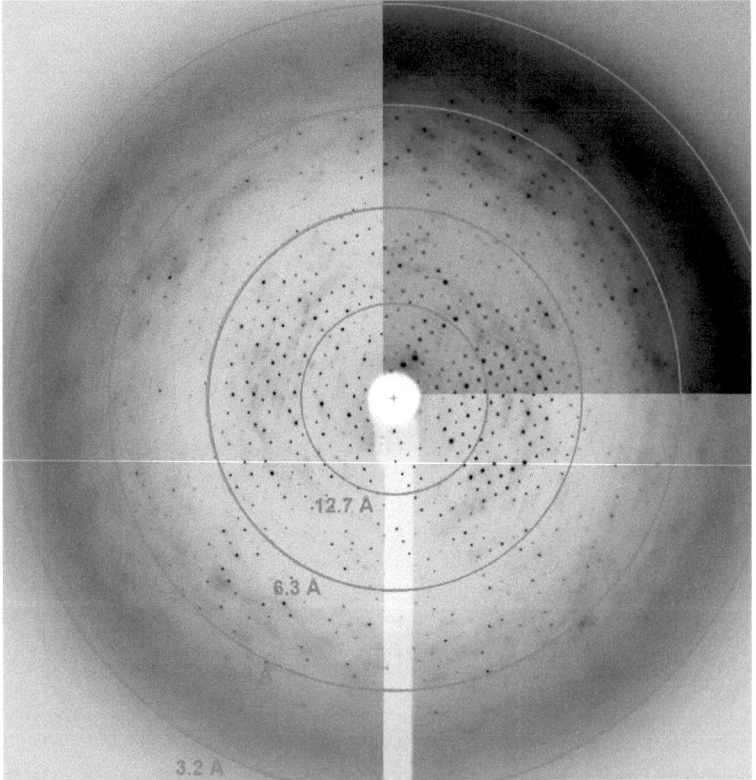

Figure 3-16 | Diffraction pattern of a EcCaiT crystal
The data set was collected on the beamline X10SA at the SLS (Villigen, Switzerland) from the triangular pyramidal EcCaiT crystal shown in Figure 3-13 **B**. Data were collected to a resolution of 3.5 Å. The upper right quarter of the diffraction pattern was adjusted to a higher contrast level to show the high-resolution diffraction spots.

The diffraction pattern shown in Figure 3-17 was collected from an StCaiT crystal. The crystal diffracted anisotropically to 3.9 Å in one direction and to 4.5 Å in the other.

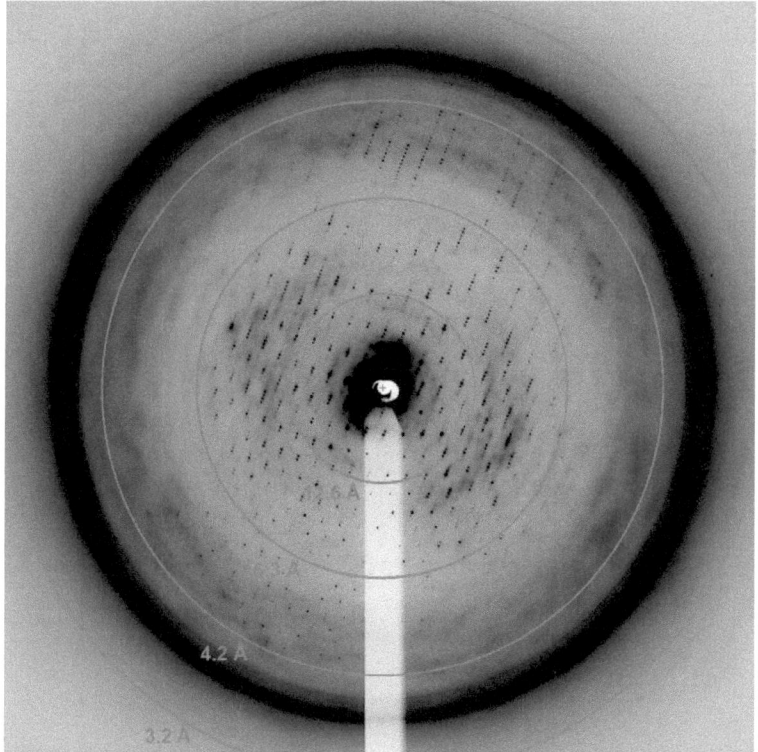

Figure 3-17 | Diffraction pattern of an StCaiT crystal

A data set was collected on the beamline ID 14-eh3 at the ESRF (Grenoble, France) from the long triangular StCaiT crystal shown in Figure 3-14 **A**. The diffraction pattern shows anisotropic diffraction to 3.9 Å in one direction but only to 4.5 Å in the other direction. Data were collected to a resolution of 3.9 Å.

Data were initially processed in $P1$ using XDS (Kabsch, 1993). The correct space group was determined using POINTLESS (Evans, 2006) and the data were scaled with SCALA (CCP4, 1994). PmCaiT crystallized in the space group $H3$ with three molecules in the asymmetric unit that assembled around a three-fold crystallographic symmetry axis (Figure 6-3). EcCaiT crystallized in $P3_2$ (Figure 6-4) and StCaiT crystallized in the space group $P6_5$ (Figure 6-5), each with three molecules in the asymmetric unit. The data collection statistics are shown in Table 3-1.

Results

Table 3-1 | Data collection statistics

	PmCaiT	EcCaiT	StCaiT
Beamline	X10SA SLS	X10SA SLS	ID-eh3 ESRF
Wavelength (Å)	0.92	0.98	0.98
Resolution (Å) [a]	19.7 – 2.3 (2.4 – 2.3)	24.7 – 3.4 (3.5 – 3.4)	24.8 – 4.2 (4.2 – 4.0)
Space group	$H3$	$P3_2$	$P6_5$
Cell dimensions	$a = b = 129.2$ Å, $c = 160.3$ Å, $\alpha = \beta = 90°, \gamma = 120°$	$a = b = 124.2$ Å, $c = 154.6$ Å, $\alpha = \beta = 90°, \gamma = 120°$	$a = b = 128.8$ Å, $c = 309.5$ Å, $\alpha = \beta = 90°, \gamma = 120°$
Number of measured reflections	246 131	136 423	382 769
Number of unique reflections	43 243	71 232	45 504
Completeness (%)	97.7 (91.2)	97.7 (87.1)	98.9 (95.5)
Redundancy	5.7	1.9	8.4
I/σI	9.0 (1.4)	12.8 (1.3)	16.8 (1.2)
R_{merge} (%) [b]	8.4 (55.7)	6.0 (90.0)	9.0 (93.4)

[a] Numbers in parentheses represent statistics for data in the highest resolution shell

[b] $R_{merge} = \dfrac{\sum_{hkl} \sum_i \| F_{hkl} | - | F_{hkl}(i) \|}{\sum_{hkl} \sum_i | F_{hkl}(i) |}$

3.3.2 Phase determination, model building and refinement

The structures of all three CaiT homologs were solved by molecular replacement (MR) using a poly-alanine model of BetP (PDB code 2WIT) as a search model. The success of this approach was surprising, as CaiT and BetP share a sequence identity of only 27% (Figure 1-8, Figure 6-2), and proved to be in different conformations. To minimize model bias during phasing, prime-and-switch phasing was performed using the program RESOLVE (Terwilliger, 2001, 2004). The CaiT model was build and refined with iterative rounds of manual building and rebuilding in COOT (Emsley and Cowtan, 2004) and O (Jones et al., 1991), followed by refinement in PHENIX (Adams et al., 2010; Adams et al., 2004; Adams et al., 2002), which modified the model to satisfy different constrains and restrains (e.g. bond angles or bond length). During refinement bulk-solvent correction, simulated annealing followed by maximum likelihood-based energy minimization and isotropic B-factor refinement was applied.

Figure 3-18 shows the initial electron density map of PmCaiT after MR (**A**), the electron density map after applying prime-and-switch phasing to minimize model bias (**B**) and the refined electron density map of PmCaiT (**C**).

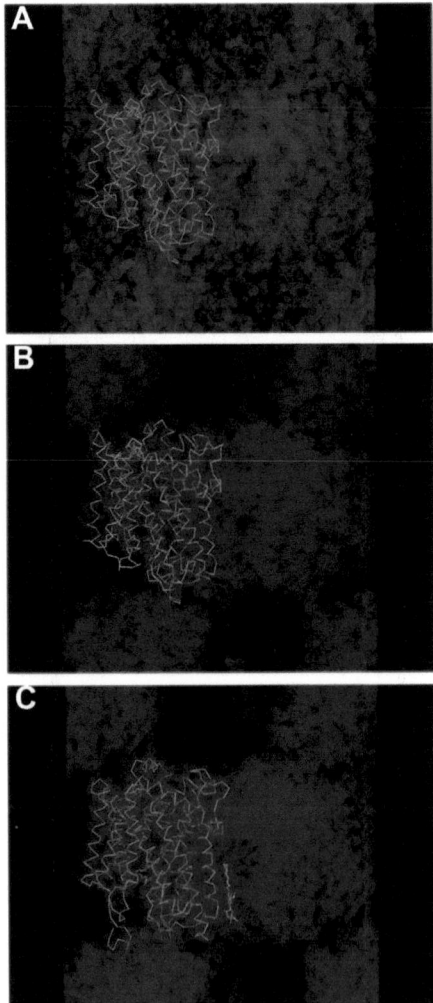

Figure 3-18 | Electron density maps and crystal packing of PmCaiT

The atomic structure of PmCaiT was determined by MR with a poly-Ala model of BetP (2WIT) as a search model. The initial electron density map after MR (**A**) was improved by applying prime-and-switch phasing that resulted in a model bias minimized electron density map (**B**). After iterative cycles of model building and rebuilding the refined electron density map (**C**) is represented by the final PmCaiT model.

Results

The refined atomic model of PmCaiT represents its final electron density map. The quality of the model is judged from the Ramachandran plot (Ramachandran *et al.*, 1963) and statistical data that compare the atomic model to the measured diffraction data (Figure 3-19). The final PmCaiT model was refined to an R_{work} of 21.2 % and an R_{free} of 23.8 %. The coordinates of the PmCaiT model and the structure factors have been deposited in the PDB with the accession code 2WSW.

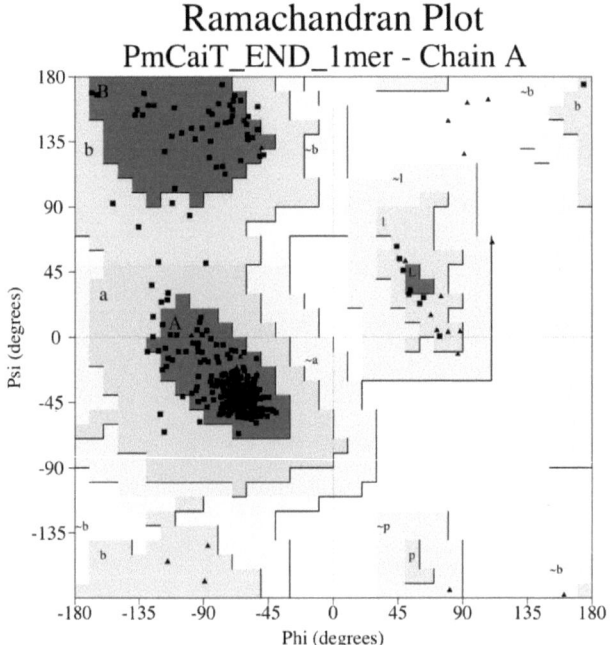

Statistics of the PmCaiT model	
PDB accession code	2WSW
Resolution (Å)	19.7 - 2.3
R_{work} (%)	20.9
R_{free} (%)	23.8
Matthews coefficient	4.46
Corresponding solvent (%)	72.23
Ramachandran plot statistics:	
Total number of residues	508
Core (%)	94.1
Allowed (%)	5.9
Generously (%)	0.0

Figure 3-19 | Ramachandran plot and statistics of the refined PmCaiT model

The Ramachandran plot shows the phi/psi torsion angles for all residues in the structure. Regions marked with a, b, l and p represent the sterically favoured φ/ψ–combinations for α–helices, β–sheets, left-handed α–helices and epsilons, respectively. The red coloured regions correspond to the most

favoured φ/ψ – combinations (core region). Allowed and generously allowed φ/ψ–combination regions are coloured in yellow and beige, respectively. Glycine residues are shown as triangles (▲) while all other residues are represented as squares (■). The Ramachandran plot of the PmCaiT model was generated using the program PROCHECK (Laskowski et al., 1993).

The structure of EcCaiT was solved by MR with PmCaiT as a poly-alanine search model. The initial model and electron density was refined as described for the refinement of PmCaiT (see above), in addition, strict non-crystallographic symmetry (NCS) restrains were enforced during refinement.

The initial electron density map of EcCaiT after MR (**A**), the electron density map with minimal model bias after applying prime-and-switch phasing (**B**) and the refined electron density map of EcCaiT (**C**) is shown in Figure 3-20.

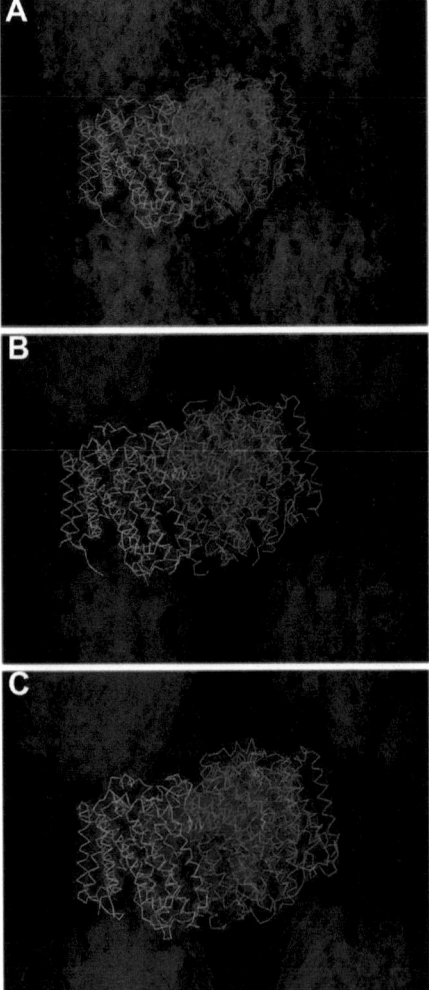

Figure 3-20 | Electron density maps and crystal packing of EcCaiT

The atomic structure of EcCaiT was determined by MR with PmCaiT as poly-Ala as a search model. The initial electron density map after MR (**A**) was improved by applying prime-and-switch phasing that resulted in an electron density map with minimal model bias (**B**). After iterative cycles of model building and rebuilding the refined electron density map (**C**) was used to build the final EcCaiT model.

The quality of the refined atomic model of EcCaiT was judged from the Ramachandran plot (Ramachandran *et al.*, 1963) and statistical data that compare the atomic model to the measured diffraction data (Figure 3-21). The final EcCaiT model was refined to an R_{work} of 23.7 % and an R_{free} of 27.1 %. The coordinates of the EcCaiT model and the structure factors have been deposited in the PDB with the accession code 2WSX.

Results

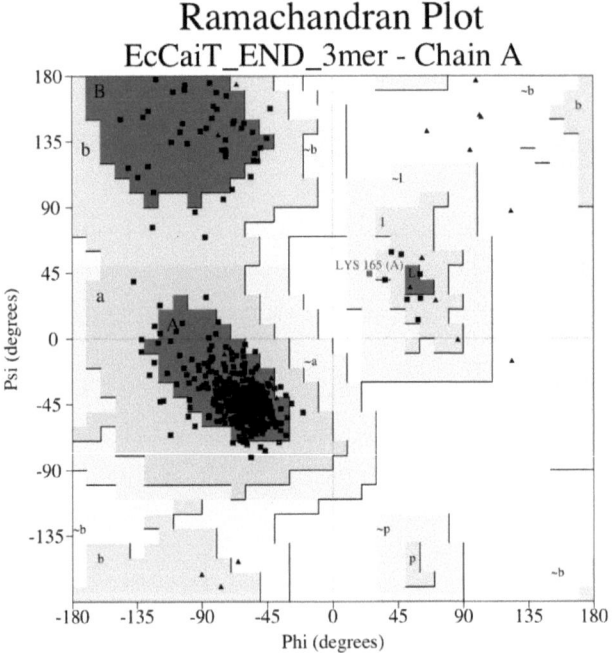

Statistics of the EcCaiT model	
PDB accession code	2WSX
Resolution (Å)	24.7 - 3.50
R_{work} (%)	23.7
R_{free} (%)	27.1
Matthews coefficient	4.18
Corresponding solvent (%)	70.35
Ramachandran plot statistics:	
Total number of residues	1488
Core (%)	90.8
Allowed (%)	9.0
Generously (%)	0.2

Figure 3-21 | Ramachandran plot and statistics of the refined EcCaiT model

The Ramachandran plot shows the phi/psi torsion angles for all residues in the structure. Regions marked with a, b, l and p represent the sterically favoured φ/ψ–combinations for α–helices, β–sheets, left-handed α–helices and epsilons, respectively. The red coloured regions correspond to the most

Results

favoured φ/ψ–combinations (core region). Allowed and generously allowed φ/ψ–combination regions are coloured in yellow and beige, respectively. Glycine residues are shown as triangles (▲) while all other residues are represented as squares (■). Lys165 is marked with a red square (■), which indicated that this residue has an unusual torsion angle and therefore was ranged into the generously allowed region. Lys165 is located in loop2 and forms in the EcCaiT crystal hydrogen bonds to the backbone carbonyl oxygens of His227 and Tyr388 of the symmetry related molecule, which causes the unusual conformation of Lys165. The Ramachandran plot of the EcCaiT model was generated using the program PROCHECK (Laskowski *et al.*, 1993).

The structure of StCaiT was solved by MR with PmCaiT as a poly-alanine search model. The initial model and electron density was refined as described for the refinement of EcCaiT (see above), except anisotropy correction and TLS refinement instead of simulated annealing was used in the later stages of refinement. At a resolution of 4 Å only a few chains are resolved. Therefore, simulated annealing at this resolution would introduce serious model bias. Anisotropy correction and TLS refinement was necessary to correct for the diffraction anisotropy. As for the refinement of the EcCaiT model, strict NCS restrains were enforced during the refinement of the StCaiT model

Figure 3-22 shows the initial electron density map of StCaiT after MR (**A**), the electron density map with minimal model bias after applying prime-and-switch phasing (**B**) and the refined electron density map of StCaiT (**C**).

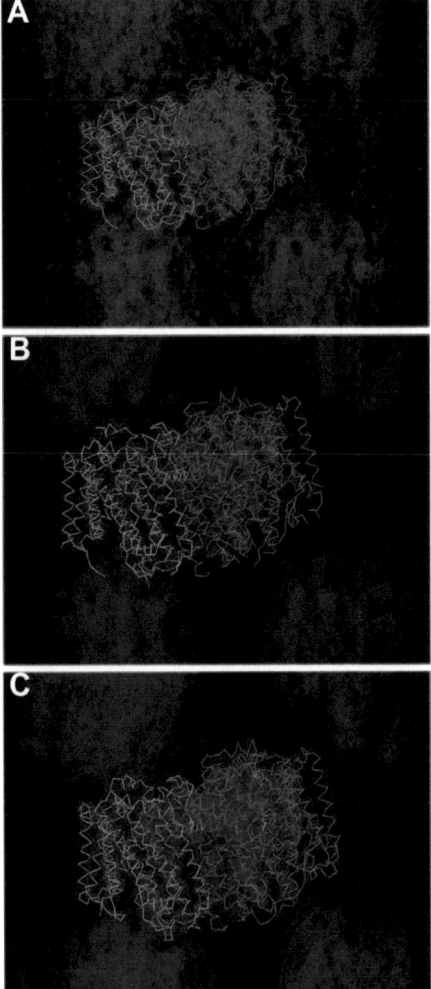

Figure 3-22 | Electron density maps and crystal packing of StCaiT

The atomic structure of StCaiT was determined by MR using PmCaiT as a poly-Ala search model. The initial electron density map after MR (**A**) was improved by applying prime-and-switch phasing that resulted in a electron density map with minimal model bias (**B**). After iterative cycles of model building and rebuilding the refined electron density map (**C**) is represented by the final StCaiT model. The packing of the protein within the unit cell $P6_5$ with the cell parameter $a = b = 128.8$ Å, $c = 309.5$ Å, $\alpha = \beta = 90°$, $\gamma = 120°$ is shown in (**D**).

Results

The quality of the refined atomic model of StCaiT was judged from the Ramachandran plot (Ramachandran et al., 1963) and statistical data that compare the atomic model to the measured diffraction data (Figure 3-23). The final StCaiT model was refined to an R_{work} of 24.2 % and an R_{free} of 27.5 %. The low R-values of the StCaiT model are due to the well-refined starting model used in MR and the use of strict three-fold NCS restraints during refinement. The coordinates and the structure factors of StCaiT have not yet been deposited in the PDB.

Results

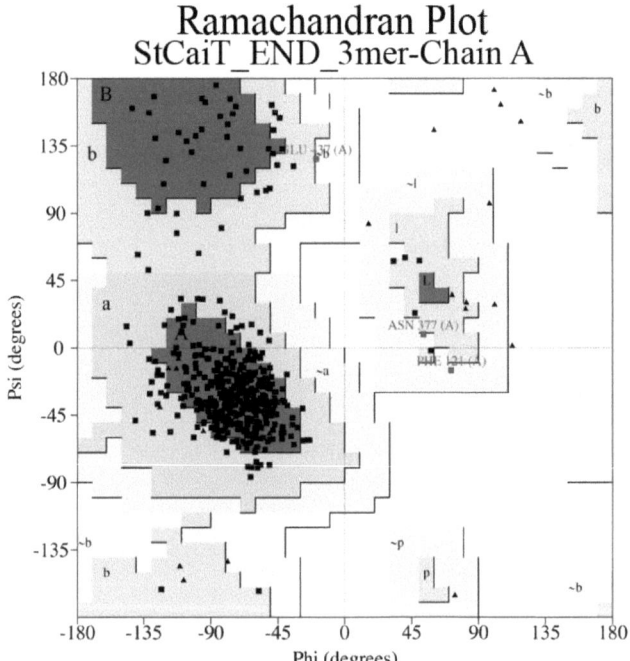

Statistics of the StCaiT model	
PDB accession code	--------
Resolution (Å)	24.8 - 4.00
R_{work} (%)	24.2
R_{free} (%)	27.5
Matthews coefficient	4.42
Corresponding solvent (%)	72.14
Ramachandran plot statistics:	
Total number of residues	1476
Core (%)	80.7
Allowed (%)	18.7
Generously (%)	0.6

Figure 3-23 | Ramachandran plot and statistics of the refined StCaiT model

The Ramachandran plot shows the phi/psi torsion angles for all residues in the structure. Regions marked with a, b, l and p represent the sterically favoured φ/ψ–combinations for α–helices, β–sheets, left-handed α–helices and epsilons, respectively. The red coloured regions correspond to the most

Results

favoured φ/ψ–combinations (core region). Allowed and generously allowed φ/ψ–combination regions are coloured in yellow and beige, respectively. Glycine residues are shown as triangles (▲) while all other residues are represented as squares (■). Residues marked with a red square (■) have unusual torsion angle and were placed into the generously allowed regions. At a resolution of 4 Å the correct assignment of residues is difficult. All these residues are located in loop regions and have less conformational restrains to other residues.

3.4 The CaiT structure

3.4.1 CaiT topology

The CaiT protomer comprises twelve transmembrane helices (TM1 to TM12), with the N- and C-termini on the cytoplasmic side as has been predicted previously (Section 1.2; (Jung et al., 2002)). The CaiT topology (Figure 3-24 **A**) resembles that of BetP (Section 1.4.1; (Ressl et al., 2009) and all its structural relatives (Abramson and Wright, 2009; Faham et al., 2008; Fang et al., 2009; Gao et al., 2009; Ressl et al., 2009; Shaffer et al., 2009; Weyand et al., 2008; Yamashita et al., 2005). CaiT containes an inverted repeat motif in which TM3 to TM7 and TM8 to TM12 form the two halves of the inverted repeat. The 5 TM helices of repeat 2 can be superimposed on the 5 TM helices of repeat 1 by ~180 ° rotation around the internal pseudo-twofold axis of the protomer, and by ~20 ° in the perpendicular direction (Figure 3-24 **B**). The inner core of the protomer is formed by TM3 and TM4 of the first repeat and TM8 and TM9 of the second repeat, which are arranged as an antiparallel four-helix bundle. TM5 to TM7 and TM10 to TM12 form a supporting framework around this inner core that is separated from the remainder of the inverted repeat by a cytoplasmic loop-helix-loop motif (L4, IH4) and a periplasmic helix-loop-helix motif (EH9a, L9, EH9b). The long, curved, amphipathic α-helix 7 (H7) runs along the membrane surface on the periplasmic side of the monomer. TM1, together with the curved TM2, form a clamp-like scaffold for the repeat helices at the back of the protomer, near the threefold axis of the trimer.

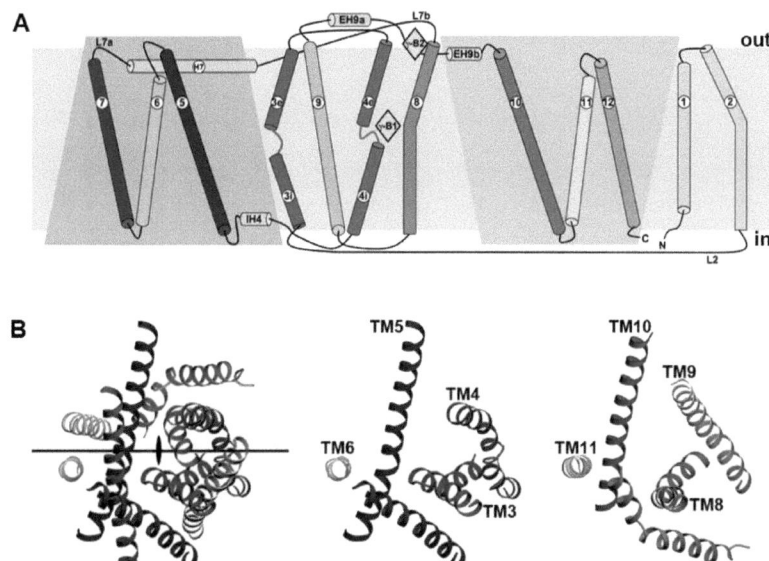

Figure 3-24 | CaiT topology of the two 5-TM inverted repeats motif

In the CaiT topology (**A**), the helices of repeat 1 (TM3 to TM7) in the inverted repeat motif are shown in dark colours, whereas the helices of repeat 2 (TM8 to TM12) are shown in lighter colours. Protein segments that do not belong to the inverted repeat motif are coloured in grey. Bound substrate is represented by yellow diamonds and is labelled with γ-B1 and γ-B2.

The 5 TM helices of the two inverted repeats (**B**) are superimposable by rotating repeat 2 (right) relative to repeat 1 (centre) by ~180 ° around the internal pseudo-twofold axis of the protomer (black line), and by ~20 ° in the perpendicular direction.

Results

3.4.2 The CaiT trimer

As expected from the high sequence identity between the three CaiT homologs (Figure 6-2), the overall structures of PmCaiT, EcCaiT and StCaiT resemble one another closely. CaiT crystallized as a symmetric trimer with three identical protomers (Figure 3-25).

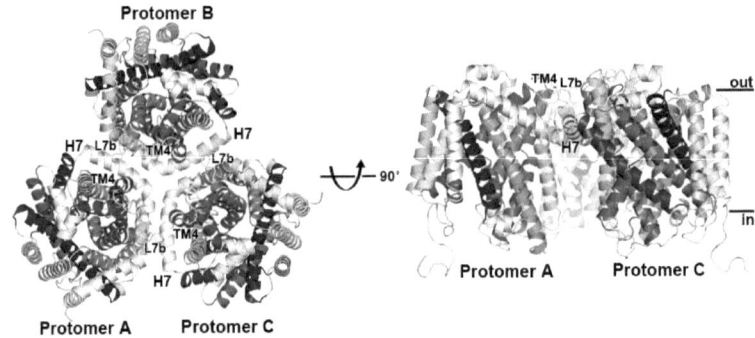

Figure 3-25 | CaiT homotrimer viewed from the periplasmic side and parallel to the membrane

Cytoplasmic (left) view and view parallel to the membrane (right) of CaiT. CaiT forms a symmetric trimer with three identical protomers. The tight interaction of the CaiT monomers (A – C) in the homotrimer are mediated by the long, curved, amphipathic helix 7 (H7), the loop connecting H7 to TM8 (L7b) and TM4. The colour code is equivalent to that of the CaiT topology.

Interactions between the protomers are mediated by the curved helix H7, by the loop connecting H7 to TM8 (L7b) and by TM4 (Figure 3-25, Figure 3-26). Arg299 in H7 forms a salt bridge to Asp288 in H7 of the neighbouring protomer. A second polar contact is the hydrogen bond that connects the hydroxyl group of Thr304 in L7b to the backbone carbonyl oxygen of Asn284 in H7 of the adjacent protomer. A water molecule forms hydrogen bonds between the carboxyl group of Asp305 and the backbone amide nitrogen of Gly308 in L7b and to the carboxyl group of Glu132 in TM4 of the next protomer (Figure 3-26). In the cavity between the three protomers detergent molecules – one per protomer – were bound (Figure 3-26). In the

Results

native lipid environment this cavity would be filled with lipid molecules, which might additionally stabilize the CaiT homotrimer.

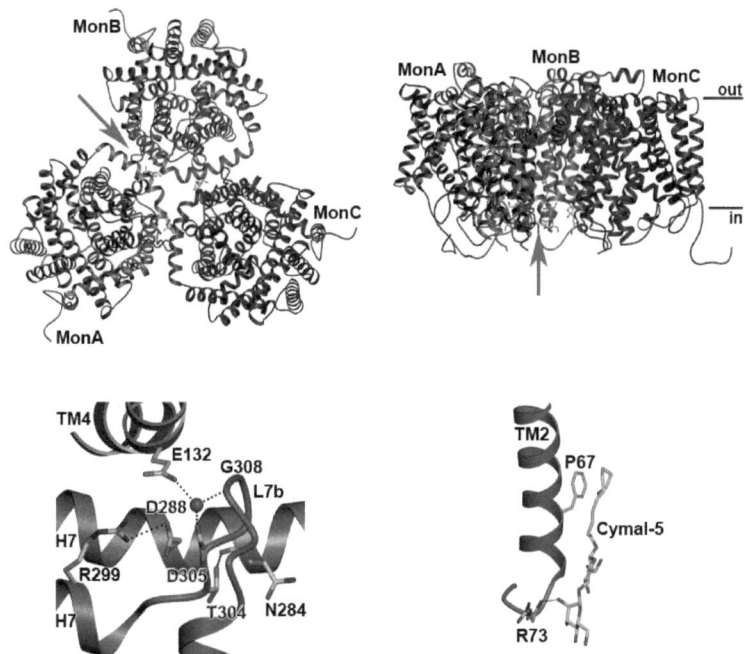

Figure 3-26 | CaiT protomer–protomer interaction and protomer–detergent interaction

Polar interactions between the CaiT protomers stabilize the homotrimer. Arg299 in H7 forms a salt bridge to Asp288 in H7 of the neighbouring protomer. A tight hydrogen bond is found between the carboxyl group of Asp305 in L7b and the backbone carbonyl oxygen of Asn284 in H7 of the next protomer. A third interaction is mediated by a water molecule (red sphere), which connects the carboxyl group of Glu132 in TM4 to the carboxyl group of Asp305 and the backbone carbonyl oxygen of Gly308 in L7b. The hydroxyl group of the maltoside in Cymal-5, which was found in the cavity between the protomers, forms a hydrogen bond to the carbonyl oxygen of Arg73 in TM2. An additional van der Waals contact between the cyclohexyl of Cymal-5 and Phe67 in TM2 holds the detergent in position.

Results

3.4.3 Cytoplasmic substrate pathway, central transport site and periplasmic substrate-binding site

In the protomer, the four-helix bundle (Figure 3-24), together with TM5, TM6 and TM11, line a ~25 Å deep funnel from the cytoplasmic membrane surface to the transport site in the centre of the protomer. The funnel is wide open on the cytoplasmic side, with a diameter of ~15 Å, and narrows to ~7 Å at the central transport site (Figure 3-27). Its entire length is unobstructed by sidechains, providing unhindered access to the central transport site. All three CaiT structures therefore clearly show the open, inward-facing conformation of the antiporter.

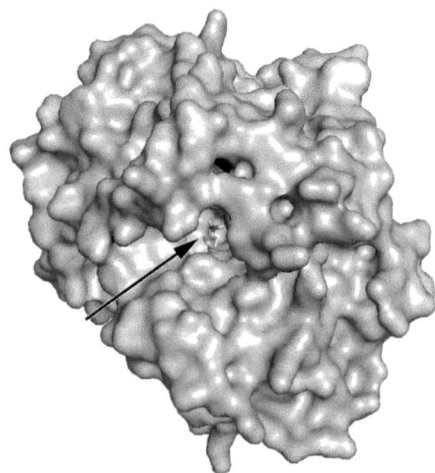

Figure 3-27 | Unobstructed cytoplasmic substrate pathway of CaiT
Cytoplasmic view down the substrate pathway to the central transport site of EcCaiT, which is occupied by γ-butyrobetaine (arrow, yellow sticks). The four-helix bundle (TM3, TM4, TM8 and TM9) together with TM5, TM6 and TM11 form a funnel that is wide open on the cytoplasmic side, with a diameter of ~15 Å, and narrows to ~7 Å at the central transport site. The entire length of the substrate pathway is unobstructed by sidechains.

Results

In the EcCaiT structure two prominent non-protein densities were found that both fitted the shape and the size of γ-butyrobetaine. One of the substrate molecules is located in the central transport site of the protomer. The second substrate molecule is bound on the extracellular surface of the protomer, at a distance of ~16 Å from the central transport site (Figure 3-28). As there was no substrate added during purification or in the crystallization trials, the molecules must have been carried over from the host cell used to express the protein.

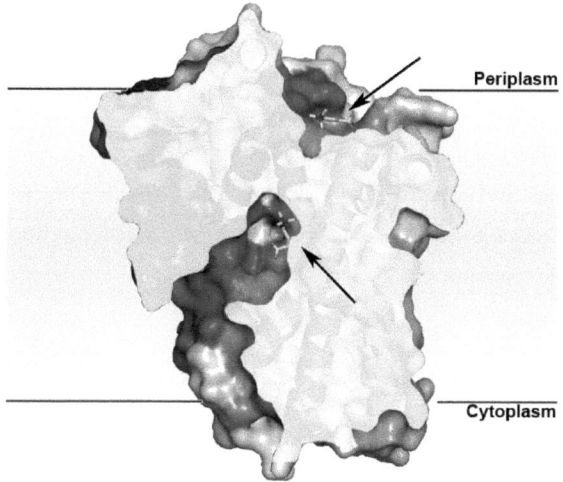

Figure 3-28 | Slice through the EcCaiT protomer, viewed parallel to the membrane

Two γ-butyrobetaine molecules are bound to the EcCaiT protomer (arrows). One of the substrate molecules is bound in the central transport site that is freely accessible from cytoplasmic side and the second is bound in a shallow cavity on the extracellular surface of the protomer.

Results

3.4.3.1 The central transport site of CaiT

The central transport site in CaiT is a tryptophan box (Figure 3-29, Figure 3-30), as in BetP (Ressl *et al.*, 2009). One half of the tryptophan box is defined by Trp142 and Trp147 from the partly unwound TM4. The other half consists of Trp323 and Trp324, which are found in the continuous but highly tilted TM8. In EcCaiT, the indole ring of Trp323 in TM8 is well placed for cation-π interaction with the positively charged ammonium group of γ-butyrobetaine (Figure 3-29).

An additional attractive interaction in the central transport site is formed by the carboxyl group of γ-butyrobetaine and the sulfur of Met331. Although the methionine side chain is usually thought of as hydrophobic, it can in fact participate in polar interactions because the large, uncharged sulfur atom is more easily polarized than smaller atoms and can therefore interact with the negatively charged carboxyl group of γ-butyrobetaine.

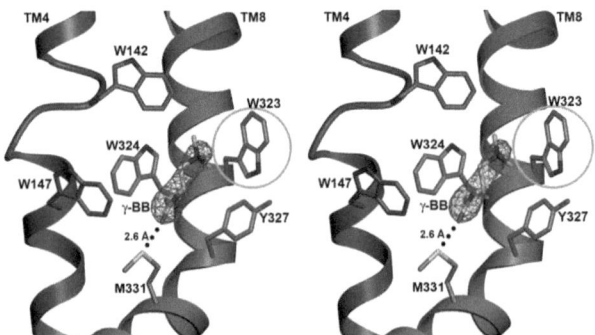

Figure 3-29 | Stereo view of the central transport site of EcCaiT with bound substrate

The central transport site is a tryptophan box that is defined by Trp142, Trp147 (TM4), Trp323 and Trp324 (TM8). The indole ring of Trp323 (light blue circle) is well placed for a cation-π interaction to the γ-butyrobetaine (γ-BB). The sulfur of Met331 coordinates the carboxyl group (dashed line) of the bound substrate by polar interaction and replaces Na1 in the Na^+-dependent transporters, accounting for Na^+ independent transport in CaiT. The $2F_o - F_c$ electron density map (blue) around the γ-butyrobetaine is contoured at 1.0 σ.

Results

The transport site of PmCaiT does not contain a bound substrate. The sidechain of Trp323 is rotated by 43° each in $\chi1$ and $\chi2$ relative to EcCaiT to a position that leaves no room for a substrate to bind (Figure 3-30 **A**). Instead the Trp box of PmCaiT contains a small solvent molecule, most likely a glycerol from the solubilizaiton buffer, which seems to form a hydrogen bond to the indole nitrogen of Trp323 (Figure 3-30 **B**). The orientation of Trp323 is the most prominent difference between the EcCaiT and PmCaiT structures.

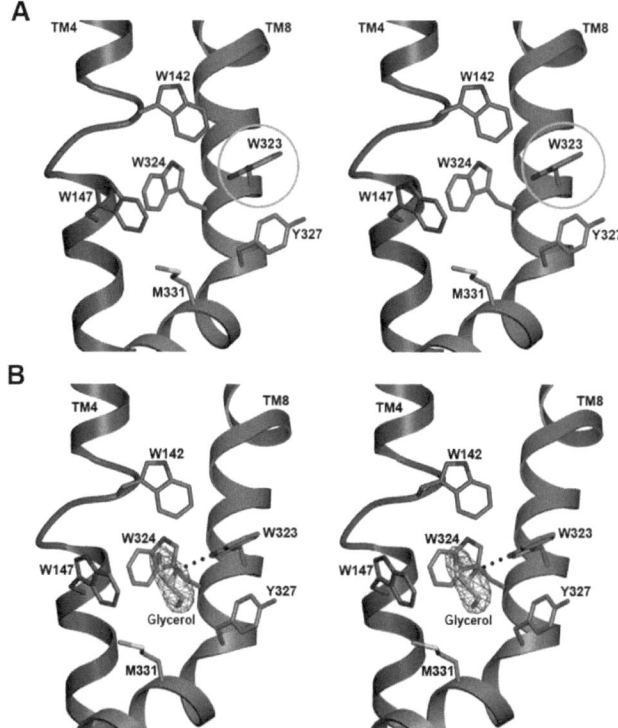

Figure 3-30 | Stereo views of the central transport site of PmCaiT

In the transport site of PmCaiT, the indole side chain of Trp323 (**A**; light blue circle) is rotated by 43° small molecule, most likely glycerol (**B**), carried over from the purification, which forms a hydrogen bond (dashed line) to the indole nitrogen of Trp323. The $2F_o - F_c$ electron density map (blue) around the glycerol molecule is contoured at 1.0 σ. The orientation of the helices in (**A**) is the same as these of EcCaiT in Figure 3-29. To show the interaction between the glycerol molecule and the indole nitrogen of Trp323, the helices in (**B**) are rotated by -17° around the y-axis.

Results

The resolution of the StCaiT structure was too low to correctly assign the orientations of the residues unambiguously. The side chains of the residues in the central transport site resemble those of PmCaiT, which is most likely due to model bias. In addition, at the resolution of 4 Å, no extra non-protein density indicating bound substrate was discernible.

3.4.3.2 The second substrate binding site of CaiT

The second substrate molecule in the extracellular cavity of EcCaiT is bound by cation-π interaction with Tyr114 in TM3 and Trp316 in TM8 and an additional hydrogen bond that is formed by the carboxyl group of γ-butyrobetaine to the backbone nitrogen of Gly315 (Figure 3-31).

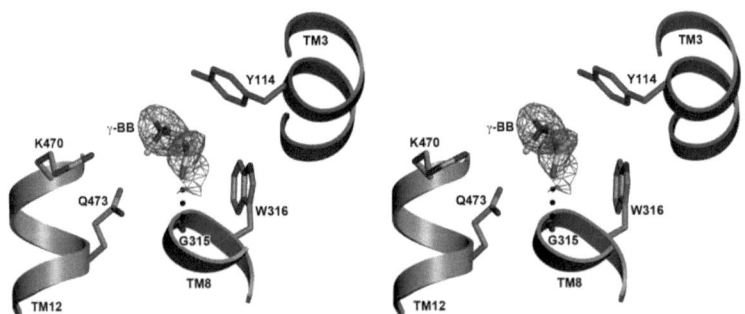

Figure 3-31 | Stereo view of the external substrate-binding site of EcCaiT
In the extracellular substrate-binding site of EcCaiT, the γ-butyrobetaine (γ-BB) is bound by cation-π interaction with Tyr114 (TM3) and Trp316 (TM8). A hydrogen bond is formed by the γ-butyrobetaine carboxyl and the backbone amide nitrogen of Gly315. The $2F_o - F_c$ electron density map (blue) around the γ-butyrobetaine is contoured at 1.0 σ.

In PmCaiT, the place of γ-butyrobetaine in the external substrate-binding site was taken by two ordered water molecules, which link Lys470 and Gln473 (TM12) to the main-chain carbonyl of Ala315 (TM8; Figure 3-31). A third resolved water

Results

molecule connects the indole nitrogen of Trp316 to the main-chain carbonyl of Tyr114 in TM3, thus establishing a hydrogen bond between the two halves of the central four-helix bundle. The same water molecule forms a hydrogen bond to the Arg(-2) sidechain of a symmetry-related protomer (Figure 3-32, Figure 3-33). Arg(-2) is part of a short linker peptide between the N-terminus and a His$_6$-affinity tag introduced for purification. This tag had been removed in EcCaiT prior to crystallization. The crystal contact blocks access to the second binding site, which might explain why the site is empty in PmCaiT.

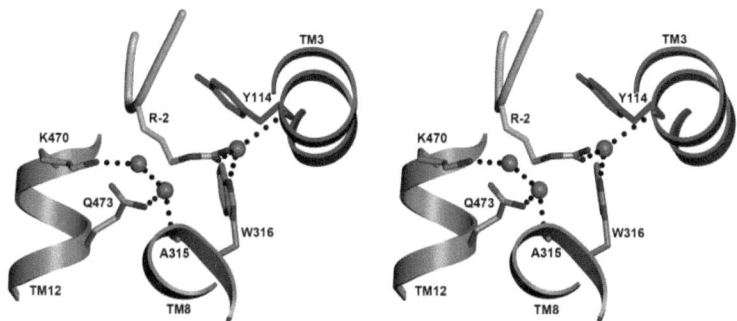

Figure 3-32 | Stereo view of the external substrate-binding site of PmCaiT

In the substrate-free external binding site of PmCaiT, two water molecules (blue spheres) mediate the interaction between TM8 to TM12. Lys470 (TM12) is connected to the backbone amide group of Ala315 (TM8) and to Gln473 (TM12) *via* two water molecules in series. Gln473 (TM12) stabilizes the interaction between TM8 and TM12 with an additional hydrogen bond. A third water molecule mediates a connection between TM3 and TM8 by forming hydrogen bonds to the carbonyl oxygen of Tyr114 (TM3) and to the indole nitrogen of Trp316 (TM8). The same water molecule forms a hydrogen bond to Arg(-2) (grey) in the linker peptide between the N-terminus and the His$_6$-affinity tag of a symmetry-related protomer. This crystal contact blocks substrate access to the external binding site.

3.4.3.3 The extracellular substrate pathway to the central binding site

Access to the central transport site from the extracellular site is blocked by bulky hydrophobic residues (Figure 3-33). On this side, Trp316 (TM8) that forms

Results

part of the external substrate-binding site, and Trp107 (TM3) protrude into the substrate pathway together with Tyr320 (TM8), which is held in a position directly above the central transport site by a hydrogen bond to the carboxyl group of Glu111 (Figure 3-33). Additional hydrophobic contacts between TM3 and TM5 of the first repeat, and TM8 and TM12 of the second repeat help to seal the pathway towards the outside.

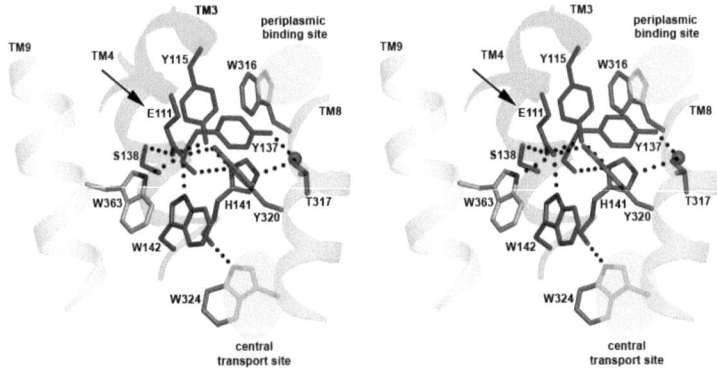

Figure 3-33 | Stereo view of the residues and hydrogen bond network blocking the substrate pathway on the periplasmic side

In the inward-facing open conformation, the extracellular substrate pathway is blocked by bulky hydrophobic residues that are held in position by forming an extensive hydrogen bond network around Glu111 (arrow). The extracellular block lines helices of the two repeats (TM3, TM4 of repeat 1 and TM8 of repeat 2). These helices must move apart when the extracellular pathway opens. The external substrate-binding site near the exit of the closed substrate pathway and the central transport site (yellow ovals) are separated by a distance of 16 Å.

3.4.3.4 Replacement of Na^+ ions in the Na^+-independent CaiT

Unlike its relative BetP, CaiT is Na^+-independent. A comparison of the structures reveals the reason for this striking difference. Although bound Na^+ ions were not resolved in the 3.35 Å BetP structure, their positions were modelled by homology to LeuT. By this model, a critical Na^+ (Na1) is coordinated by the carboxyl

Results

group of the glycine betaine substrate and the backbone carbonyl oxygens of Ala148 and Met150 in TM3 (Ressl *et al.*, 2009). Although this interaction would be equally possible in CaiT, neither the EcCaiT nor the PmCaiT structure shows an apparent Na^+ density in this region. Instead, inspection of the residues surrounding the central transport site indicates an attractive non-bonding interaction between the carboxyl group of γ-butyrobetaine and the sulfur of Met331 (Figure 3-29).

Met331 is conserved in prokaryotic CaiT and mammalian OCTN2, but not in BetP or BetT, which both require Na^+ (Figure 6-2). In BetP, the position of Met331 is taken by Val381 (Figure 1-8, Figure 6-2). Remarkably, the interaction with the Met sulfur in CaiT thus replaces the Na^+ ion coordination of the substrate in BetP.

A second well-resolved Na^+ ion (Na2) in LeuT is positioned between helices 1 and 8 (Yamashita *et al.*, 2005). This Na2 site was modelled between TM3 and TM10 in BetP (Figure 1-13; (Ressl *et al.*, 2009)). Na2 stabilizes the unwound stretch of helix 1, which in LeuT participates in substrate coordination. In CaiT, the position corresponding to Na2 is occupied by Arg262 in TM7 (Figure 3-34), which is accessible to solvent in the cytoplasmic funnel. The positive charge of Arg262 links TM3 to TM10, and stabilizes the unwound region of TM3, which contributes to substrate binding in the external binding site through Tyr114 (Figure 3-31).

Results

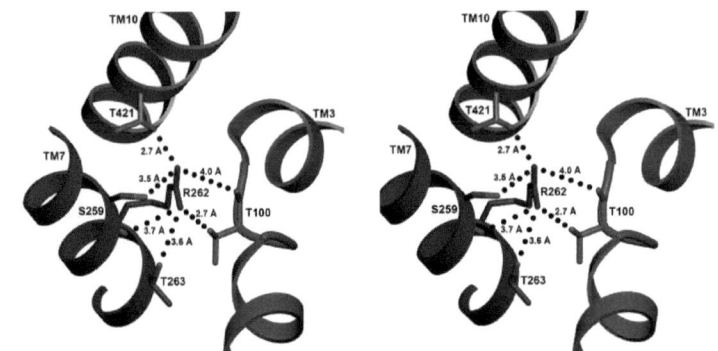

Figure 3-34 | Stereo view of the hydrogen bond network around Arg262
Arg262 (TM7) is located between TM3 and TM10 and stabilizes the unwound region of TM3 by hydrogen bonds. Thus the positive charge of Arg262 takes over the role of Na2 inLeuT and related Na^+-dependent transporters.

3.5 Lipid analysis and reconstitution of CaiT

3.5.1 Lipid analysis

Detergent solubilization and purification of membrane proteins destroys the lipid environment that surrounds the protein in the cell. Despite this harsh treatment, some lipids remain attached to the purified protein. These bound lipids often play an important role in the stabilization or functionality of the protein. Knowing the lipids that remain attached to the protein after solubilization and purification helps to define the conditions for reconstituting the active protein into liposomes which is often essential for functional studies. In addition, it is interesting to see if there is a difference in the composition of bound lipid composition between the heterologously expressed proteins PmCaiT and StCaiT and the homologoulsy expressed EcCaiT.

The lipid composition of the purified protein was evaluated by thin-layer chromatography (TLC). Silica gel TLC plates were loaded with *E. coli* lipid standards (EPL, PE and PG), a detergent (Cymal-5) standard and the purified protein sample.

In all three cases the lipid PE was observed in the protein sample (Figure 3-35) indicating that PE remained attached to CaiT after solubilization and purification. There is no obvious difference in the lipid composition between the homologously and heterologously expressed CaiT proteins.

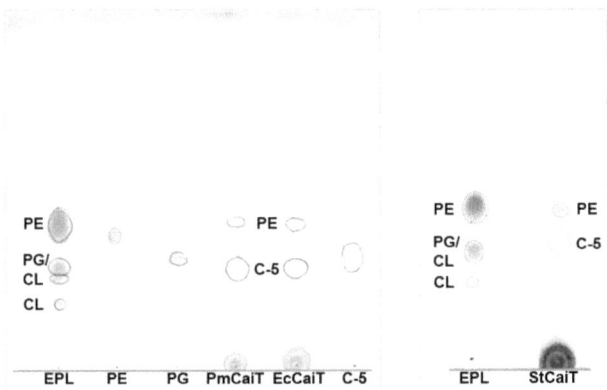

Figure 3-35 | Silica gel TLC plate of *E. coli* lipids, Cymal-5, PmCaiT, EcCaiT and StCaiT

The samples were loaded on a TLC plate and the lipids were separated using a chloroform : methanol : water (69 % : 27 % : 4 % v/v) mixture. The dried plate was first stained with iodine vapour to visualize electron rich molecule moieties like carbon-carbon double bonds of unsaturated fatty acids or the maltoside of detergent molecules like Cymal-5 or DDM. Afterwards the plate was stained with molybdenum spray reagent to detect phospholipids. The commercially available *E. coli* lipids *E. coli* polar lipids (**EPL**), Phosphatidylethanolamine (**PE**) and Phosphatidylglycerol (**PG**) were used as lipid standards and Cymal-5 (**C-5**) as detergent standard. 50 – 100 µg of the purified protein samples (**PmCaiT, EcCaiT** and **StCaiT**) were loaded onto the TLC plate.

3.5.2 Reconstitution of CaiT into liposomes

Activity studies were performed using CaiT reconstituted into *E. coli* polar lipids (EPL). The reconstitution was first tested with EcCaiT and a lipid-to-protein ratio (LPR) of 20:1. The resulting proteoliposomes as assessed by freeze facture (Figure 3-36) showed evenly distributed proteins in liposomes that measured between 400 – 600 nm in diameter. These proteoliposomes were suitable for binding and transport studies.

Results

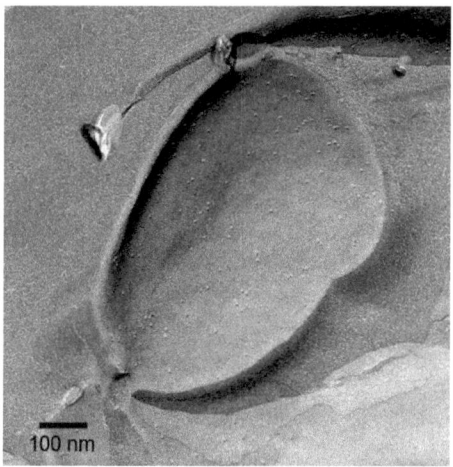

Figure 3-36 | Freeze-fracture image of EcCaiT wt reconstituted into EPL liposomes
Reconstitution was achieved with EPL liposomes and purified EcCaiT wt at an LPR of 20:1. The proteoliposomes were extruded first through a 200 nm membrane filter followed by extrusion through a 400 nm membrane filter.

Initial binding and transport measurements on proteoliposomes with an LPR of 20:1 showed promising results but the signal was to weak for detailed analysis, an LPR of 10:1 was required. All binding and transport measurements with reconstituted EcCaiT wt, PmCaiT wt and PmCaiT mutants (Figure 3-37), and StCaiT wt (Figure 3-38) were performed with proteoliposomes of an LPR of 10:1.

Results

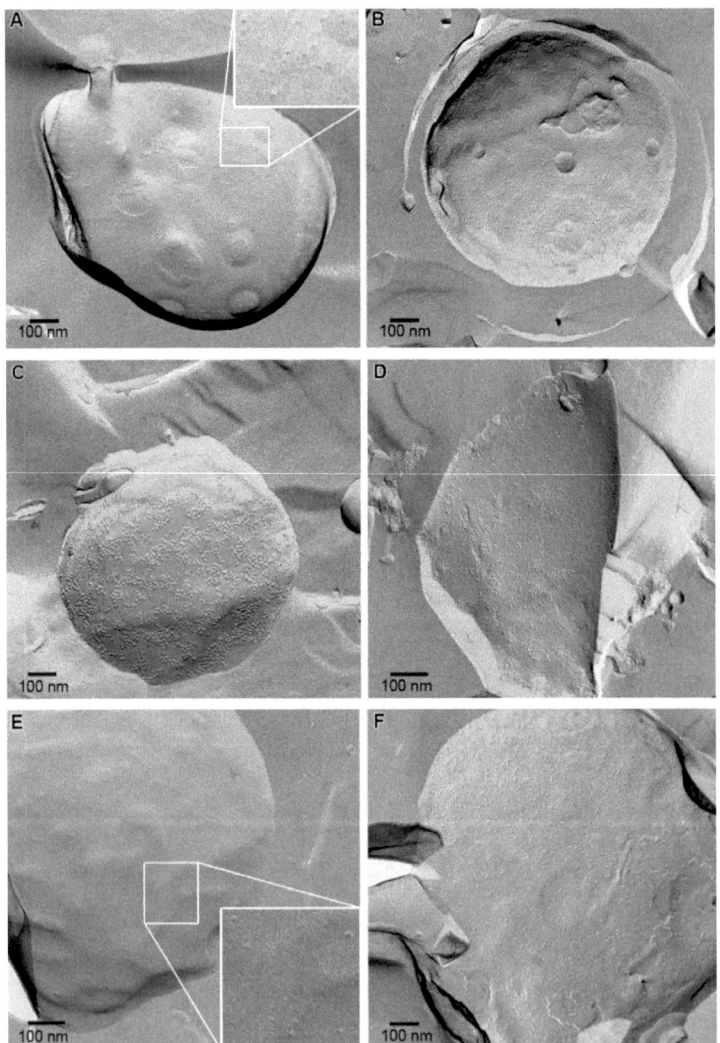

Figure 3-37 | Freeze-fracture images of PmCaiT wt and PmCaiT mutants reconstituted into EPL liposomes

Proteoliposomes used for binding and transport studies were checked by freeze-fracture (performed by F. Joos, MPI of Biophysics). Reconstitution of PmCaiT wt was first tested with an LPR of 20:1 (**A**). PmCaiT wt proteoliposomes at an LPR of 10:1 (**B**) were more suitable for binding and transport studies. In (**C**) – (**F**) the mutants PmCaiT W316A, PmCaiT M331V, PmCaiT E111A and PmCaiT R262E were reconstituted with an LPR of 10:1.

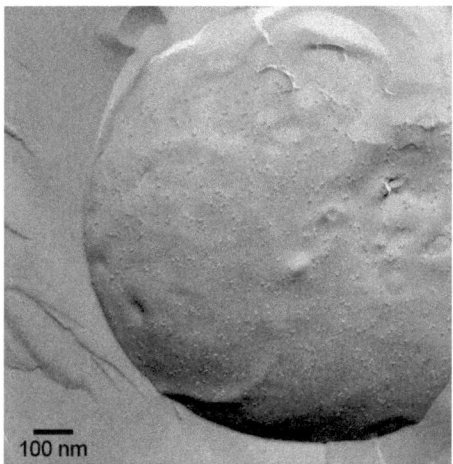

Figure 3-38 | Freeze-fracture image of StCaiT wt reconstituted into EPL–liposomes

StCaiT wt was reconstituted into EPL–liposomes at an LPR of 10:1.

Results

3.6 Substrate transport and binding studies

3.6.1 Substrate transport studies

3.6.1.1 Wildtype CaiT

To study substrate transport by the three homologs PmCaiT, EcCaiT and StCaiT, the protein had to be reconstituted into liposomes. The proteoliposomes (Figure 3-36, Figure 3-37, Figure 3-38) were preloaded with a saturating concentration (10 mM) of non-radioactive potential substrates such as glycine, glycine betaine, L-carnitine and γ-butyrobetaine. The impermeability of the proteoliposomes was controlled with unloaded proteoliposomes. Substrate exchange was measured by recording the uptake of ^{14}C-labelled L-carnitine. Plotting the initial rates of L-carnitine uptake against the external concentration of ^{14}C-labelled L-carnitine resulted in a saturation curve for proteoliposomes that were preloaded either with non-radioactive L-carnitine or γ-butyrobetaine (Figure 3-39). CaiT clearly discriminates between its natural substrates, L-carnitine and γ-butyrobetaine, and other potential substrates like glycine or glycine betaine. The data were fitted to the Michaelis-Menten equation as previously described (Jung *et al.*, 2002) and values for the binding constant (K_M) and the maximum substrate uptake (v_{max}) were calculated using the program Origin 7.5. The kinetic analysis of CaiT protoliposomes preloaded with L-carnitine gave a K_M for L-carnitine of 40 – 90 µM and a v_{max} of 3000 – 5200 nmol L-carnitine/(min•mg protein) (Table 3-2) for all three homologs. Analysis of CaiT proteoliposomes preloaded with γ-butyrobetaine gave a K_M for L-carnitine of 45 – 100 µM and a v_{max} of 4600 – 5800 nmol L-carnitine/(min•mg protein) (Table 3-2) for the three homologs.

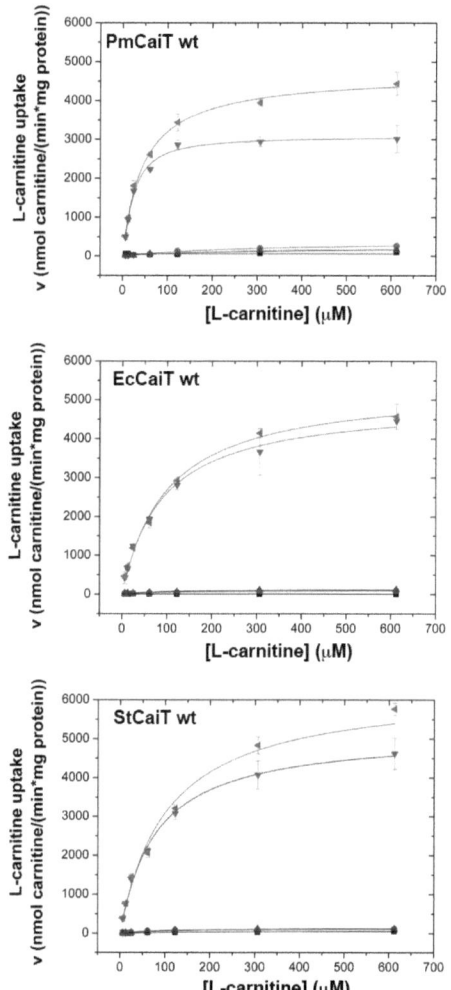

Figure 3-39 | L-carnitine uptake into proteoliposomes with reconstituted PmCaiT, EcCaiT or StCaiT

CaiT was reconstituted into liposomes at an LPR of 10 and buffered in sodium-free Tris buffer. The resulting proteoliposomes were preloaded with non-radioactive potential substrates such as glycine (●), glycine betaine (▲), L-carnitine (▼) and γ-butyrobetaine (◄). The impermeability of the proteoliposomes was tested with unloaded proteoliposomes as a control (■). Substrate uptake was performed in sodium-free buffer and all measurments were repeated 3 – 5 times.

Results

The sodium dependence of PmCaiT and EcCaiT uptake was tested with protein reconstituted into liposomes buffered in Tris with 100 mM NaCl. The proteoliposomes were preloaded with γ-butyrobetaine and ^{14}C-labelled L-carnitine uptake was recorded (Figure 3-40). Data analysis indicated a K_M of 100 – 120 µM and a v_{max} of 4800 – 5000 nmol L-carnitine/(min•mg protein) (Table 3-2) for PmCaiT and EcCaiT with NaCl present. Comparison of the parameters with the kinetic parameters calculated from measurements under sodium-free conditions show that the K_M increases slightly in the presence of NaCl. The maximum substrate uptake (v_{max}) does not change when NaCl is added.

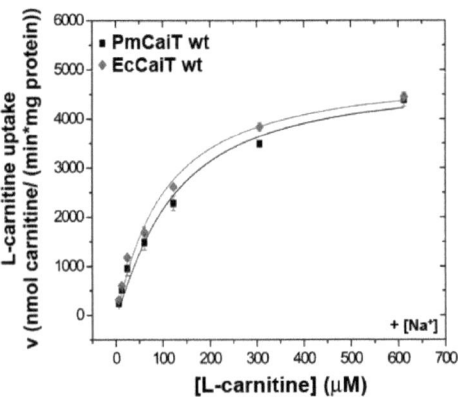

Figure 3-40 | L-carnitine uptake into proteoliposomes with reconstituted PmCaiT or EcCaiT in the presence of Na$^+$

CaiT was reconstituted into liposomes at an LPR of 10 and buffered in Tris with 100 mM NaCl. The proteoliposomes were preloaded with γ-butyrobetaine and ^{14}C-labelled L-carnitine uptake was measured for PmCaiT (■) and EcCaiT (♦) in the presence of 100 mM NaCl.

Table 3-2 gives an overview of the substrate transport parameters for wildtype PmCaiT, EcCaiT and StCaiT.

Table 3-2 | **Kinetic analysis of PmCaiT wt, EcCaiT wt and StCaiT wt**

	PmCaiT wt	EcCaiT wt	StCaiT wt
proteoliposomes preloaded with L-carnitine			
K_M (µM)	43 ± 2	89 ± 10	79 ± 6
v_{max} (nmol L-carnitine/ (min•mg protein))	3081 ± 98	4932 ± 186	5158 ± 132
k_{cat} (L-carnitine/min)	173 ± 6	280 ± 11	292 ± 7
proteoliposomes preloaded with γ-butyrobetaine			
K_M (µM)	46 ± 6 119.7 ± 20 [a]	82 ± 12 100 ± 14 [a]	103 ± 9
v_{max} (nmol L-carnitine/ (min•mg protein))	4672 ± 205 4824 ± 301 [a]	4921 ± 243 4975 ± 243 [a]	5735 ± 368
k_{cat} (L-carnitine/min)	263 ± 12 272 ± 17 [a]	279 ± 14 282 ± 14 [a]	325 ± 21

[a] Measurements were performed in buffers containing 100 mM NaCl

3.6.1.2 PmCaiT mutants

Based on the CaiT structure, different mutants were designed and their influence on substrate uptake was tested. The following mutants were of special interest: Met331 which mimics the Na1 in the central transport site, Trp316 that contributes to the coordination of the external substrate, Glu111 which is the central residue in the external substrate pathway that holds the bulky residues in position to close the pathway in the inward-facing open conformation, and Arg262 that mimics the Na2 and stabilizes the unwound region of TM3.

The mutation of Met331 to valine shows a drastical reduced transport activity compared to wildtype PmCaiT (Figure 3-41). The kinetic analysis of PmCaiT Met331Val proteoliposomes preloaded with γ-butyrobetaine gave a K_M for L-carnitine of 120 µM and a v_{max} of 480 nmol L-carnitine/(min•mg protein) (Table 3-3). The binding affinity of the mutant is reduced by a factor of 2 while the substrate

Results

uptake is reduced by a factor of 10 compared to PmCaiT wt (Figure 3-41, Table 3-2, Table 3-3). Met331 in CaiT mimicks Na1 of LeuT (Figure 1-12) and BetP (Figure 1-13) and therefore has a crucial role during substrate transport while substrate binding is less affected by the mutation as it is mainly enabled by the tryptophans of the tryptophan box.

The mutation of Trp316 to alanine reduces the transport activity by a factor of 3 compared to PmCaiT wt (Figure 3-41). The K_M of Trp316Ala for L-carnitine was calculated as 160 µM and v_{max} as 1800 nmol L-carnitine/(min•mg protein) (Table 3-3). The L-carnitine uptake of PmCaiT Trp316Ala was measured with proteoliposomes that were preloaded with γ-butyrobetaine.

The mutation of Glu111 to alanine renders the protein inactive (Figure 3-41). Replacing Glu111 by Ala not only effects the proteins transport activity but also de-stabilizes the trimer (Figure 3-42), highlighting the crucial role of this residue for CaiT assembly and function.

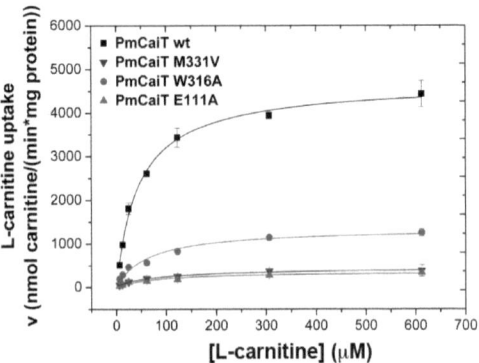

Figure 3-41 | Substrate transport activity of PmCaiT mutants
The PmCaiT mutants were reconstituted into liposomes with an LPR of 10 and buffered in sodium-free Tris buffer. The proteoliposomes were preloaded with γ-butyrobetaine and ^{14}C-labelled L-carnitine uptake was measured for PmCaiT wt (■), PmCaiT Met331Val (▼), PmCaiT Trp316Ala (●) and PmCaiT Glu111Ala (▲).

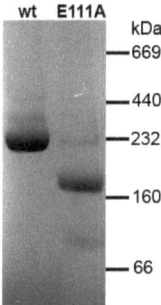

Figure 3-42 | BN-PAGE gradient gel of PmCaiT E111A and PmCaiT wt

Wildtype PmCaiT (**wt**) forms stable trimers that run at about 232 kDa in the BN-PAGE gradient gel. Replacing Glu111 by alanine (**E111A**) destabilizes the trimer and results in the appearance of dimers and monomers.

The mutant in which Arg262 is changed to glutamate shows a striking substrate transport behaviour. The mutant renders the protein inactive when no sodium is present but transport activity is partially rescued by Na^+ (Figure 3-43). The kinetic analysis of PmCaiT Arg262Glu proteoliposomes preloaded with γ-butyrobetaine gave a K_M for L-carnitine of 20 µM and a v_{max} of 1300 nmol L-carnitine/(min•mg protein) (Table 3-3) in the presence of 100 mM NaCl. Compared to PmCaiT wt the binding constant of PmCaiT Arg262Glu is reduced by a factor of 2.5 while the maximum substrate uptake is reduced by a factor of 3.6.

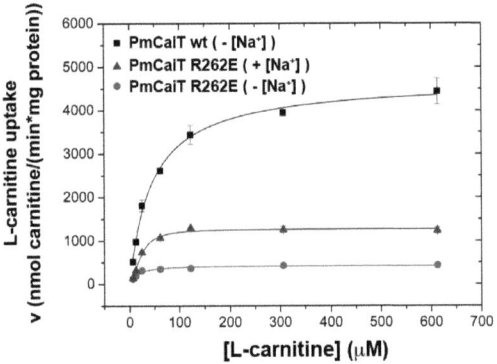

Figure 3-43 | Na$^+$-dependent substrate transport of PmCaiT Arg262Glu
L-carnitine uptake by PmCaiT wt (■) and PmCaiT Arg262Glu with (▲) and without (●) 100 mM NaCl. Na$^+$ can partially rescue the transport activity of Arg262Glu mutant.

Table 3-3 summarizes the transport activity parameters of the PmCaiT mutants Glu111Ala, Arg262Glu, Trp316Ala and Met331Val.

Table 3-3 | Substrate transport activity constants of PmCaiT mutants

	PmCaiT E111A	PmCaiT R262E without NaCl	PmCaiT R262E with NaCl	PmCaiT W316A	PmCaiT M331V
proteoliposomes preloaded with γ-butyrobetaine					
K_M (µM)	inactive	inactive	21 ± 2	159.2 ± 11	123 ± 12
v_{max} (nmol L-carnitine/ (min•mg protein))	inactive	inactive	1272 ± 32	1816 ± 353	478 ± 53
k_{cat} (L-carnitine/min)	inactive	inactive	72 ± 2	69 ± 4	26 ± 3

Results

3.6.2 Binding studies

Substrate binding to CaiT is readily monitored by tryptophan fluorescence, which senses the chemical environment of tryptophan sidechains. Comparisons of the PmCaiT and EcCaiT structures show that the main difference between the two is the orientation and the chemical environment of Trp323. These changes are induced upon substrate binding that can be monitored by tryptophan fluorescence measurements.

3.6.2.1 Binding studies of CaiT in detergent solution

3.6.2.1.1 Wildtype CaiT

To determine the substrate binding parameters of PmCaiT, EcCaiT and StCaiT, the intrinsic fluorescence changes of the homologs were first tested with different potential substrates such as glycine, glycine betaine, L-carnitine, γ-butyrobetaine and choline (Figure 3-44). The intrinsic fluorescence of the protein without substrate was recorded for reference.

The potential substrates glycine and choline had no effect on the intrinsic fluorescence of CaiT. The addition of glycine betaine at saturation concentration (25 mM) influences the intrinsic fluorescence of CaiT. This is reflected by an increase in the fluorescence intensity in all three homologs. The increase in fluorescence was more prominent for EcCaiT and StCaiT (~25 %) than for PmCaiT (~15 %) (Figure 3-44).

The addition of the CaiT substrates L-carnitine or γ-butyrobetaine shows a clear effect on tryptophan fluorescence. The substrates induce changes in both the fluorescence intensity and the emission maximum (red shift; Figure 3-44) reflecting substrate binding. At saturating concentrations of L-carnitine or γ-butyrobetaine the fluorescence intensity increases by 30 – 40 % and the emission maximum shifts from 338 nm to 342 nm (Figure 3-44).

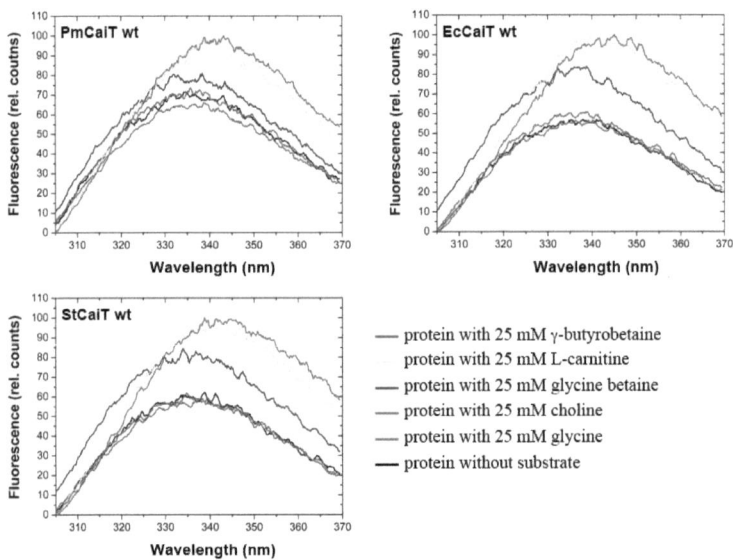

Figure 3-44 | Influence of potential substrates on the intrinsic fluorescence of PmCaiT wt, EcCaiT wt and StCaiT wt

Fluorescence spectra of PmCaiT wt (top left), EcCaiT wt (top right) and StCaiT wt (lower left) show the effect of different potential substrates on tryptophan fluorescence. The protein was excited at 295 nm (slit width 2.5 nm) and fluorescence emission was recorded from 305 nm to 370 nm (slit width 5.0 nm). All binding measurements were repeated 3 – 5 times.

To calculate binding parameters for all three CaiT homologs the concentration-dependent fluorescence change was tested for γ-butyrobetaine (Figure 3-45) and L-carnitine (Figure 3-46) in the concentration range from 0.5 mM to 100 mM. The fluorescence spectra show that the protein is saturated at a substrate concentration of ~25 mM. At a substrate concentration of 0.5 mM, the fluorescence intensity increase is not obvious but the red shift is detectable. For all subsequent binding assays the fluorescence change was recorded in the substrate concentration

Results

range of 0.25 mM to 40 mM. The intrinsic fluorescence of protein without substrate was recorded for reference.

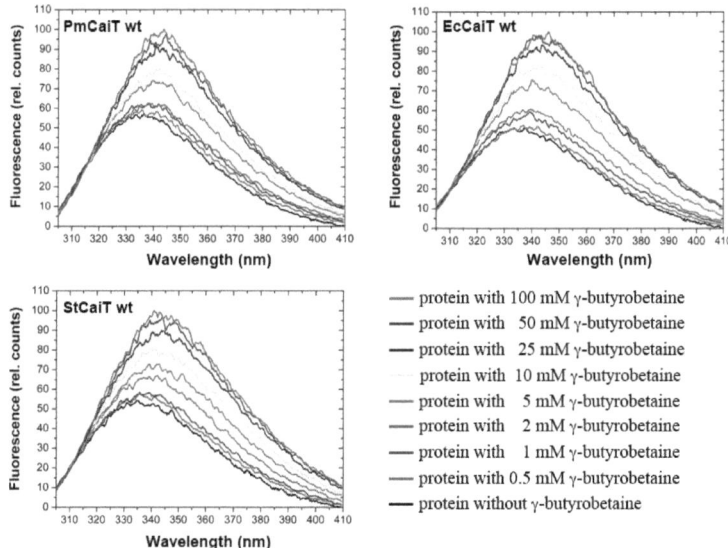

Figure 3-45 | γ-butyrobetaine induced fluorescence change of PmCaiT wt, EcCaiT wt and StCaiT wt

Fluorescence spectra of PmCaiT wt (top left), EcCaiT wt (top right) and StCaiT wt (lower left) show the concentration-dependent effect of the natural substrate γ-butyrobetaine on the CaiT tryptophan fluorescence.

Results

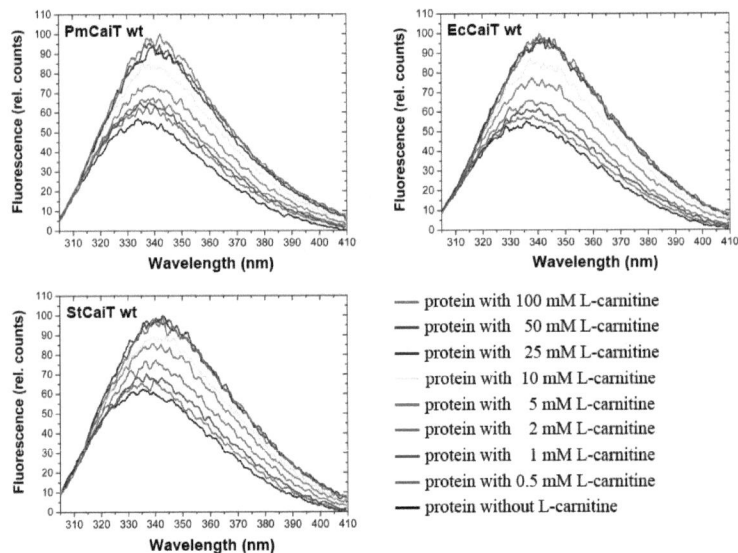

Figure 3-46 | L-carnitine induced fluorescence change of PmCaiT wt, EcCaiT wt and StCaiT wt

Fluorescence spectra of PmCaiT wt (top left), EcCaiT wt (top right) and StCaiT wt (lower left) show the concentration-dependent effect of the natural substrate L-carnitine on the CaiT tryptophan fluorescence.

Plotting the fluorescence change (ΔF/F) at a wavelength of 342 nm against the concentration of titrated substrate resulted in a saturation curve (Figure 3-47). The data were fitted to the Michaelis-Menten equation as previously described (Jung *et al.*, 2002) and values for the dissociation constant (K_D), the maximum substrate binding concentration (B_{max}) and the Hill coefficient (h) were calculated using the program Origin 7.5. The substrate binding analysis of CaiT solubilized in Cymal-5 gave a K_D for γ-butyrobetaine of 2.9 – 3.9 mM, a B_{max} of 40 – 43 mM and a Hill coefficient of 1.0 – 1.2 for the three CaiT homologs (Table 3-4). The respective analysis for PmCaiT wt, EcCaiT wt or StCaiT wt with L-carnitine gave a K_D of 1.8 – 1.9 mM, a B_{max} of 39 – 43 mM and a Hill coefficient of 1.0 – 1.2 (Table 3-4).

Results

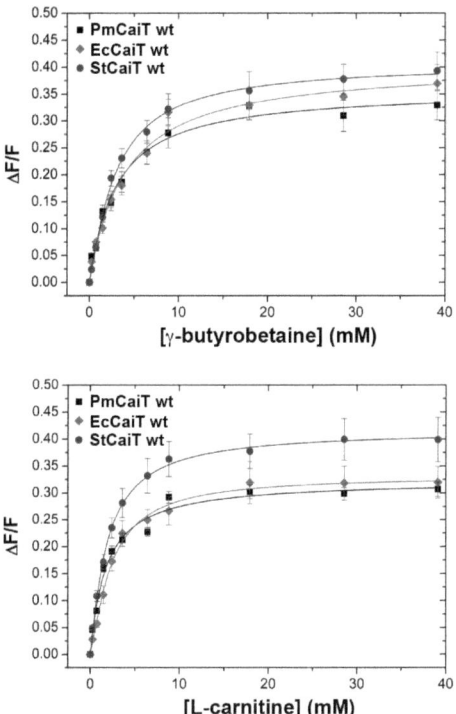

Figure 3-47 | Binding of γ-butyrobetaine or L-carnitine to PmCaiT wt, EcCaiT wt or StCaiT wt, monitored by tryptophan fluorescence

Binding of γ-butyrobetaine (top diagram) or L-carnitine (lower diagram) to PmCaiT (■), EcCaiT (♦) or StCaiT (●) solubilized in Cymal-5 resulted in a saturation curve from which the binding constants for each substrate can be calculated. The measurements were performed with sodium-free Tris buffer and the proteins were purified without NaCl in sodium-free Tris buffer.

For PmCaiT wt the influence of Na^+ or H^+ on substrate binding was tested (Figure 3-48). The binding assays were performed in the presence of different concentrations of Na^+ or at pH 7.0 and pH 9.0. At pH 5.0 the protein aggregated and binding studies were not possible.

The substrate binding analysis for L-carnitine with 50 mM NaCl gave a K_D of 2.4 ± 0.4 mM, a B_{max} of 39 ± 2.3 mM and a Hill coefficient of 1.0 ± 0.1. The respective analysis for PmCaiT wt with L-carnitine at pH 9.0 gave a K_D of 3.1 ± 0.3 mM, a B_{max} of 40.4 ± 2.6 mM and a Hill coefficient of 1.2 ± 0.1 (Table 3-4).

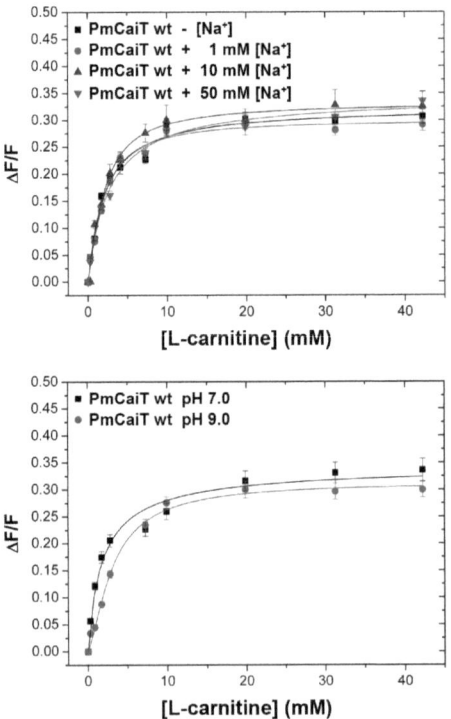

Figure 3-48 | Na^+- or pH-independent L-carnitine binding of PmCaiT wt
Influence of the L-carnitine binding to PmCaiT wt solubilized in Cymal-5 under different Na^+ concentrations (top diagram) or at different pH (lower diagram). In the left diagram L-carnitine binding to PmCaiT wt was tested in the presence of no NaCl (■), 1 mM NaCl (●), 10 mM NaCl (▲) or 50 mM NaCl (▼). The right diagram shows L-carnitine binding to PmCaiT wt at pH 7.0 (■) or pH 9.0 (●). None of the tested parameters had an influence on the substrate binding to CaiT.

Table 3-4 summarizes the substrate binding parameters for the wildtype CaiT homologs PmCaiT, EcCaiT and StCaiT.

Table 3-4 | Substrate binding constants for PmCaiT wt, EcCaiT wt and StCaiT wt

	PmCaiT wt without NaCl	PmCaiT wt with NaCl	EcCaiT wt	StCaiT wt
K_D (γ-butyrobetaine) (mM)	3.1 ± 0.6	3.4 ± 0.2	3.9 ± 0.7	2.9 ± 0.2
B_{max} (γ-butyrobetaine) (mM)	42.0 ± 1.0	42.0 ± 1.0	43.3 ± 3.2	40.8 ± 2.0
Hill coefficient (γ-butyrobetaine)	1.0 ± 0.1	1.1 ± 0.1	1.0 ± 0.1	1.2 ± 0.1
K_D (L-carnitine) (mM)	1.8 ± 0.3 3.1 ± 0.3 [a]	2.4 ± 0.4	1.9 ± 0.2	1.9 ± 0.1
B_{max} (L-carnitine) (mM)	40.3 ± 2.5 40.4 ± 2.6 [a]	39.1 ± 2.3	39.1 ± 3.1	43.2 ± 3.6
Hill coefficient (L-carnitine)	1.1 ± 0.1 1.2 ± 0.2 [a]	1.0 ± 0.1	1.2 ± 0.1	1.1 ± 0.1

[a] Measurements were performed at pH 9.0

3.6.2.1.2 PmCaiT mutants

For a deeper understanding of the CaiT exchange mechanism, the binding behaviour of two PmCaiT mutants were studied. The question was how mutations of Met331 that coordinates the substrate in the central transport site, or of Trp316 that contributes to the coordination of the external substrate, would affect CaiT function.

The binding of γ-butyrobetaine or L-carnitine to Met331Val was measured in the absence or presence of NaCl (Figure 3-49). The substrate binding analysis gave a K_D for γ-butyrobetaine of ~12 mM in the absence or presence of NaCl (Table 3-5). Compared to PmCaiT wt (Table 3-4) the binding constant of the mutant to γ-butyrobetaine is reduced by a factor of 4. The K_D for L-carnitine of the PmCaiT Met331Val was ~7 mM without or with NaCl (Table 3-5). This is again a reduction of the binding affinity by a factor of 4 compared to PmCaiT wt (Table 3-4). The analysis gave a B_{max} of 42 – 51 mM and a Hill coefficient of 1.0 – 1.2 for both

Results

substrates γ-butyrobetaine and L-carnitine and in absence or presence of NaCl (Table 3-5).

The binding analysis for the mutant Trp316Ala gave a K_D of 17.5 ± 6.0 mM for γ-butyrobetaine and 9.0 ± 1.3 mM for L-carnitine, which when compared to PmCaiT wt (Table 3-4) equals a reduction in the binding affinity for γ-butyrobetaine or L-carnitine by a factor of 6 or 5, respectively. The analysis gave a B_{max} of 58.6 ± 7.8 mM for γ-butyrobetaine and 33.7 ± 1.3 mM for L-carnitine and a Hill coefficient of 1.0 for both substrates (Table 3-5).

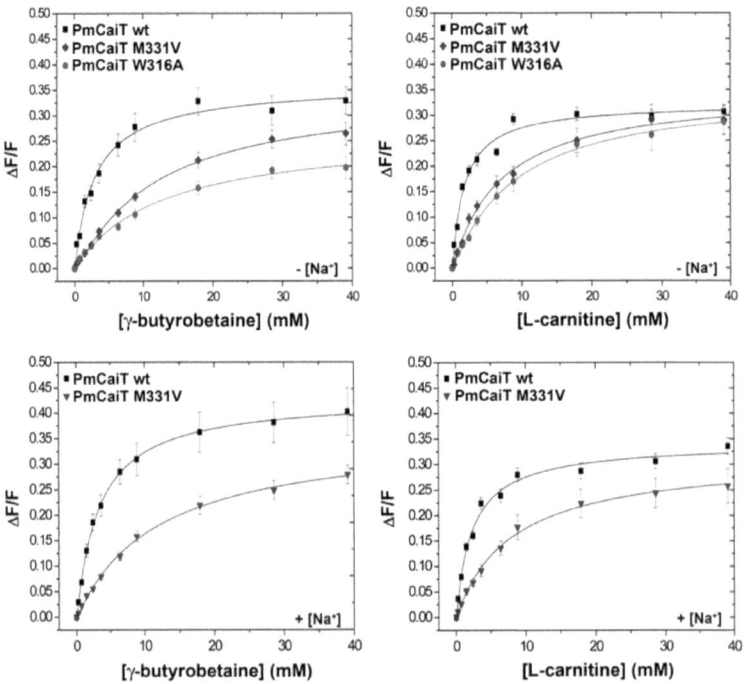

Figure 3-49 | Substrate binding activity of PmCaiT M331V and PmCaiT W316A

Binding of γ-butyobetaine and L-carnitine to PmCaiT wt (■), PmCaiT Met331Val (♦,▼) and PmCaiT W316 (●) solubilized in Cymal-5 in sodium-free Tris buffer (top) or in Tris buffer containing 50 mM NaCl.

Table 3-5 gives an overview of the substrate binding parameters of PmCaiT Met331Val and PmCaiT Trp316Ala.

Table 3-5 | Substrate binding constants of PmCaiT M331V and PmCaiT W316A

	PmCaiT M331V without NaCl	PmCaiT M331V with NaCl	PmCaiT W316A
K_D (γ-butyrobetaine) (mM)	11.7 ± 1.2	11.8 ± 1.7	17.5 ± 6.0
B_{max} (γ-butyrobetaine) (mM)	51.1 ± 3.0)	50.2 ± 2.8	58.6 ± 7.8
Hill coefficient (γ-butyrobetaine)	1.1 ± 0.1	1.2 ± 0.1	0.9 ± 0.1
K_D (L-carnitine) (mM)	7.1 ± 1.1	7.0 ± 0.8	9.0 ± 1.3
B_{max} (L-carnitine) (mM)	41.7 ± 2.7	45.1 ± 1.5	33.7 ± 1.3
Hill coefficient (L-carnitine)	1.1 ± 0.1	1.0 ± 0.1	1.0 ± 0.1

3.6.2.2 Binding studies of reconstituted CaiT

In addition to the substrate binding studies with CaiT solubilized in Cymal-5, measurements with reconstituted PmCaiT and EcCaiT were performed. The binding characteristics of the protein might be different in the natural lipid environment. The mutant PmCaiT Trp316Ala was considered in these measurements because the question was if the mutation of the external substrate-binding site has an effect on the transport mechanism which might be more pronounced under natural conditions.

The substrate binding curves for CaiT reconstituted into liposomes are sigmoidal (Figure 3-50), indicating positive cooperative substrate binding. The cooperativity is not apparent with detergent-solubilized CaiT (Figure 3-47, Figure 3-48, Figure 3-49).

The binding analysis gave a K_D for γ-butyrobetaine of 5.6 ± 0.8 mM or 5.3 ± 0.8 mM and a B_{max} of 34.3 ± 2.9 mM and 21.8 ± 1.8 mM for PmCaiT wt and EcCaiT

Results

wt, respectively (Table 3-6). The Hill coefficient for the PmCaiT wt and EcCaiT wt is 1.5 ± 0.1 and 1.4 ± 0.1, respectively.

The analysis for L-carnitine as substrate gave a K_D of 4.5 ± 0.8 mM and 5.8 ± 1.0 mM and a B_{max} of 28.3 ± 1.9 mM or 30.5 ± 1.5 mM for PmCaiT wt or EcCaiT wt (Table 3-6). The calculated Hill coefficients for PmCaiT wt and EcCaiT wt are 1.4 ± 0.1 and 1.5 ± 0.1, respectively.

Both the binding affinity and the maximal substrate binding concentration for the two substrates are reduced when the protein is reconstituted into liposomes (Table 3-4) compared to the detergent-solubilized wildtype CaiT (Table 3-4).

Binding analysis for the reconstituted Trp316Ala mutant resulted in K_D values for γ-butyrobetaine or L-carnitine of 11.2 ± 1.7 mM or 7.5 ± 1.2 mM (Table 3-6). These values are comparable with the substrate affinity constants for PmCaiT Trp316Ala solubilized in Cymal-5 (Table 3-5). The B_{max} values 21.8 ± 1.8 mM or 24.4 ± 0.8 mM for γ-butyrobetaine or L-carnitine are reduced compared to the detergent-solubilized PmCaiT Trp316Ala (Table 3-5). This result is similar to that seen for the wildtype CaiTs that was reconstituted into liposomes. The Hill coefficient for reconstituted PmCaiT Trp316Ala is 1.7 ± 0.1 for γ-butyrobetaine and 1.6 ± 0.1 for L-carnitine (Table 3-6), indicating the presence of pronounced cooperativity for the binding of γ-butyrobetaine and L-carnitine in proteoliposomes.

Results

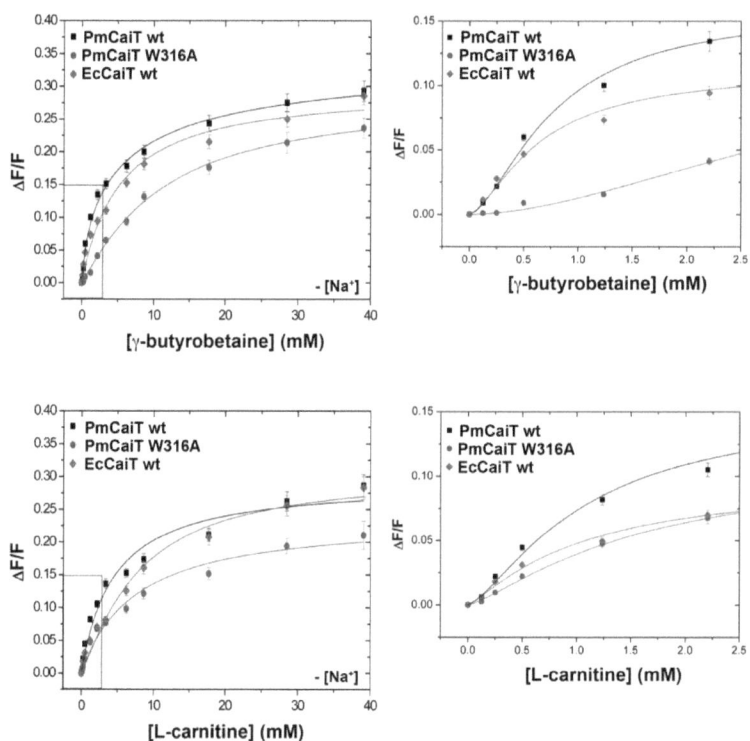

Figure 3-50 | Binding studies of reconstituted PmCaiT wt, PmCaiT Trp316Ala and EcCaiT wt

PmCaiT wt, PmCaiT Trp316Ala or EcCaiT was reconstituted into liposomes at an LPR of 10, buffered in sodium-free Tris. The proteoliposomes were titrated with γ-butyrobetaine (top diagram) or L-carnitine (lower diagram). Substrate binding for PmCaiT wt (■), PmCaiT Trp316A (●) and EcCaiT (♦) was monitored by tryptophan fluorescence. The protein was excited at 295 nm (slit width 2.5 nm) and fluorescence emission was monitored at 342 nm (slit width 5.0 nm).

The calculated substrate binding parameters for CaiT reconstituted into liposomes are summarized in Table 3-6.

Table 3-6 | Substrate binding constants of PmCaiT wt, PmCaiT W316A and EcCaiT wt reconstituted into liposomes

	PmCaiT wt	PmCaiT W316A	EcCaiT wt
K_D (γ-butyrobetaine) (mM)	5.6 ± 0.8	11.2 ± 1.7	5.3 ± 0.8
B_{max} (γ-butyrobetaine) (mM)	34.3 ± 2.9	21.8 ± 1.8	22.8 ± 0.8
Hill coefficient (γ-butyrobetaine)	1.5 ± 0.1	1.7 ± 0.2	1.4 ± 0.1
K_D (L-carnitine) (mM)	4.5 ± 0.8	7.5 ± 1.2	5.8 ± 1.0
B_{max} (L-carnitine) (mM)	28.3 ± 1.9	24.4 ± 0.8	30.5 ± 1.5
Hill coefficient (L-carnitine)	1.4 ± 0.1	1.6 ± 0.1	1.5 ± 0.2

4 Discussion

In this work the 3D X-ray structures of the secondary substrate/product antiporter CaiT from three different species were solved. To obtain more information about the substrate transport mechanism of CaiT, binding and transport analyses were performed on the wildtype protein and on mutants that were designed on the basis of the crystal structure.

4.1 Crystallization behaviour of the three CaiT homologs

The EcCaiT project was taken over from K. R. Vinothkumar, who had obtained EcCaiT 2D crystals (Figure 1-6) that grew in the presence of DPPC (dipalmitoylphosphatidylcholine) and POPC (palmitoyl-oleyl-*sn*-phophatidylcholine). The projection map (Figure 1-6) obtained from negatively stained vesicular two-dimensional crystals of EcCaiT had a resolution of 25 Å and showed $P3$ symmetry with a unit cell dimension of $a = b = 93$ Å and $\gamma = 120$ ° (Figure 1-6; (Vinothkumar *et al.*, 2006)). K. R. Vinothkumar also started 3D crystallization of CaiT which yielded crystals that diffracted to ~6 Å (Vinothkumar, 2005).

The CaiT homologs of *P. mirabilis* and *S. typhimurium* were chosen out of 388 homologs after analyzing the results of the crystallization feasibility calculated by the *XtalPred server* (http://ffas.burnham.org/XtalPred). All 388 CaiT homologs were categorized as "very difficult". The selection of the two CaiT homologs used for crystallization was therefore based on two criteria: the isoelectric point (pI) and the hydrophobicity index or gravy index. Since EcCaiT has a pI of 8.54 (Figure 3-7) one homolog was chosen with a lower and the other with a higher pI value. PmCaiT has a pI of 8.41 and StCaiT has a pI of 8.67 (Figure 3-8, Figure 3-9). Comparisons of pI values from crystallized proteins that were deposited in data bases (Section 2.4.3.1) showed that the pI of 8.41 from PmCaiT lies in a pI value region of "observed

Discussion

successes" (Figure 3-8). The pI value of 8.67 from StCaiT lies in a pI value region of "observed failure" (Figure 3-9).

The gravy index gives an indication of the hydrophobicity of a protein and thus how likely the protein is able to establish crystal contacts between molecules. The optimal gravy index lies in the region of -0.4 – 0.2. Membrane proteins tend to have a high gravy index since a considerable part of the protein has to interact with the hydrophobic acyl chains of the lipids in the membrane. The hydrophobicity index of most CaiT homologs is 0.72 as it is for EcCaiT (Figure 3-7) and StCaiT (Figure 3-9). One of the few exceptions is PmCaiT, which has a gravy index of 0.68 (Figure 3-8). The lower gravy index for PmCaiT suggests that it may crystallize more easily.

The crystallization behaviour of the three CaiT homologs was quite different from the initial crystallization trials. Both EcCaiT and PmCaiT formed sharp edged rhomboidal or rectangular crystals whereas StCaiT crystallized in small needles. The crystallization prediction calculated by the *XtalPred server* correlates well with the real crystallization behaviour of the CaiT homologs. PmCaiT had the best crystallization criteria in the prediction with the pI value in the "observed success" range and the lowest gravy index. The diffraction data collected from PmCaiT crystals was used to solve the structure of CaiT.

The gravy indices of EcCaiT and StCaiT are identical but the pI value of EcCaiT is 0.13 units lower than that of StCaiT and was categorized in the "observed success" region (Figure 3-7). EcCaiT crystallized better than StCaiT but not as well as PmCaiT. Although diffraction data was collected for phasing from EcCaiT cyrstals, the structure could not be solved without the well-refined PmCaiT model.

StCaiT has a pI of 8.67 and thus was categorized in the "observed failure" region (Figure 3-9). Compared to the other two CaiT homologs, StCaiT crystals were difficult to optimize and even the optimized crystals were fragile. Due to the low resolution, the StCaiT structure is biased by the PmCaiT model.

The crystal contact observed in PmCaiT is mainly formed by Arg(-2) that is located in the thrombin cleavage site between the protein and the His$_6$-tag and residues forming in the external substrate binding site of CaiT (Figure 3-32). When the linker peptide is included in the calculation for the crystallization feasibility the gravy index for PmCaiT decreases to 0.67 (Figure 6-7), while the pI increases to 8.56 (Figure 6-7), which is still categorized in the "observed success" region. All three homologs were crystallized with and without His$_6$-tag and therefore with and without the linker peptide. PmCaiT only crystallized well with the His$_6$-tag left at the protein while EcCaiT and StCaiT crystallized better with the His$_6$-tag cleaved off. EcCaiT formed crystal contacts between residues in loop L9b and residues in the loops L2 and L10 of the symmetry related protomer. StCaiT formed crystal contacts by residues in the loops L2 and L10 with residues in the loops L3 and L9b of the symmetry related protomer.

Although PmCaiT is clearly an unexpected positive exception, the crystallization tendency of the three CaiT homologs followed the *XtalPred server* predictions. However, all parameters analyzed in these predictions are based on soluble proteins. A similar prediction server for membrane proteins is needed which estimates, apart from the stability and tertiary structure of the protein of interest, the charge distribution of the extracellular and intracellular surface at certain pH values, determines the hydrophobic region and possible hydrophobic interactions and estimates the length and charge distribution of loop regions also at different pH values.

4.2 Oligomeric state of CaiT

The oligomeric state of CaiT has been analyzed by K. R. Vinothkumar, who showed using different methods that CaiT forms stable trimers in detergent solution and in the lipid membrane of 2D cyrstals (Vinothkumar *et al.*, 2006). In single particle analysis of CaiT it was shown that in detergent solution two CaiT trimers tend to stack together and form hexamers (Figure 4-1; (Vinothkumar *et al.*, 2006)). The hexamer formation was also observed during BN gradient gel electrophoresis (Figure 3-5). When equal concentrations of the three homologs were loaded onto a BN gradient gel, EcCaiT and StCaiT usually separated into two bands corresponding to a CaiT trimer and hexamer (Figure 3-5). PmCaiT separated into three (Figure 3-5)

Discussion

or four bands, which shows that PmCaiT has a high tendency to associate into higher oligomers without forming unordered aggregates – a behaviour that is advantageous for 3D crystallization.

When the structure of the CaiT trimer is modeled into the side view oriented 2D average of single particles (Figure 4-1 B), an interaction between the α-helical loop L9b and loop L2 of the next protomer is possible. A similar contact is observed in the crystal structures of EcCaiT and StCaiT. It is likely that this interaction is the reason for the formation of CaiT hexamers in detergent solution. The tendency of PmCaiT to form higher oligomers can be explained by the strong interaction of the N-terminus and residues in L9b, TM8 and TM3.

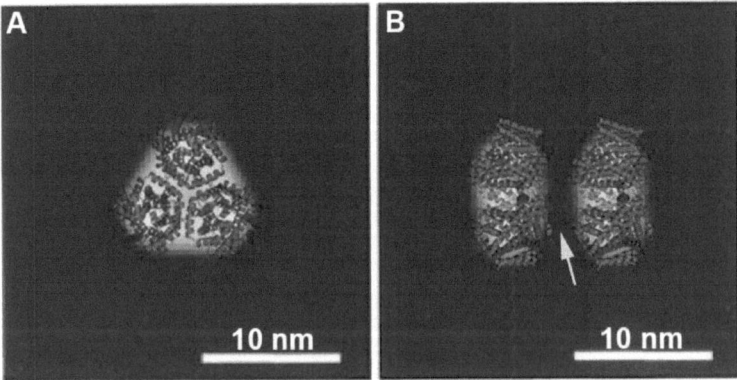

Figure 4-1 | 3D CaiT structure modeled into 2D averages of single particles

The 3D CaiT structure is modeled into the face-on view (**A**) and side view (**B**) of 2D averaged of CaiT single particles (Vinothkumar *et al.*, 2006). In the side view two CaiT trimers interact. The distance between the two trimers is about 6 Å (yellow arrow), side chains not included. This distance allows interactions between residues in the helical formed loop L9b and loop L2 of the next protomer.

Discussion

4.3 The CaiT structure

4.3.1 CaiT trimer architecture

CaiT forms compact trimers of three identical protomers (Figure 4-2). The three protomers are arranged in a closely packed triangle with a length of about 100 Å in the longest dimension. The surface potential of each protomer of CaiT (Figure 4-2) shows a predominantly negatively charged periplasmic surface and a predominantly positively charged cytoplasmic surface. This electrostatic distribution is in agreement with the positive-inside rule (von Heijne and Gavel, 1988). In addition, both the N- and C-terminus, which are located in the cytoplasm, are positively charged. The distance between the two surfaces measures ~45 Å and consists of non-polar residues that are engaged in protein-lipid interactions. The width of the hydropobic part is in good agreement with the experimentally determined thickness of pure phophatidylcholine (PC) bilayers with 18:1 acyl chains, where the phosphodiester groups are 38 Å apart and the hydrophobic core is 27 Å thick (Lewis and Engelman, 1983; Palsdottir and Hunte, 2004). Although PC is not a naturally compound of the *E. coli* membrane, the experimentally determined thickness of the PC bilayer is comparable to the native membrane in *E. coli* (Lewis and Engelman, 1983; Palsdottir and Hunte, 2004). The three protomers within the CaiT trimer form a hydrophobic cavity (Figure 4-2) on the cytoplasmic side that measures about 24 Å in height and about 15 – 20 Å in diameter. In the cavity of PmCaiT, three Cymal-5 detergent molecules were found, each biniding to one protomer (Figure 3-26). These detergent molecules are likely to be replaced by lipids in the native membrane, which would additionally stabilize the trimer. Lipid analysis of the purified protein show that PE remains bound to the protein after purification (Figure 3-35), but no density could be assigned to a lipid molecule without ambiguity.

Strong ionic and polar interactions between the protomers (Section 3.4.1; Figure 3-26) ensure trimer formation and make the trimer more rigid than that of BetP, where interactions mediated by residues in H7 are mainly hydrophobic (Ressl

Discussion

et al., 2009) and comparatively weak. In BetP the flexibility within the trimer is thought to enable the protomers to activate their neighbours (Ressl *et al.*, 2009) whereas the rigid trimer architecture of CaiT suggests that each protomer functions independently similar as the glutamate transporter from *Pyrococcus horikoshii* Glt$_{Ph}$ (Reyes *et al.*, 2009; Yernool *et al.*, 2004), which also crystallized as rigid trimers (Yernool *et al.*, 2004). However, the transport domains are expected to move stochastically and independently within each trimer (Reyes *et al.*, 2009; Yernool *et al.*, 2004) and crosslinking studies lead to the conclusion that the crystal structure represents the rare case when all three protomers are trapped in the same conformation (Reyes *et al.*, 2009).

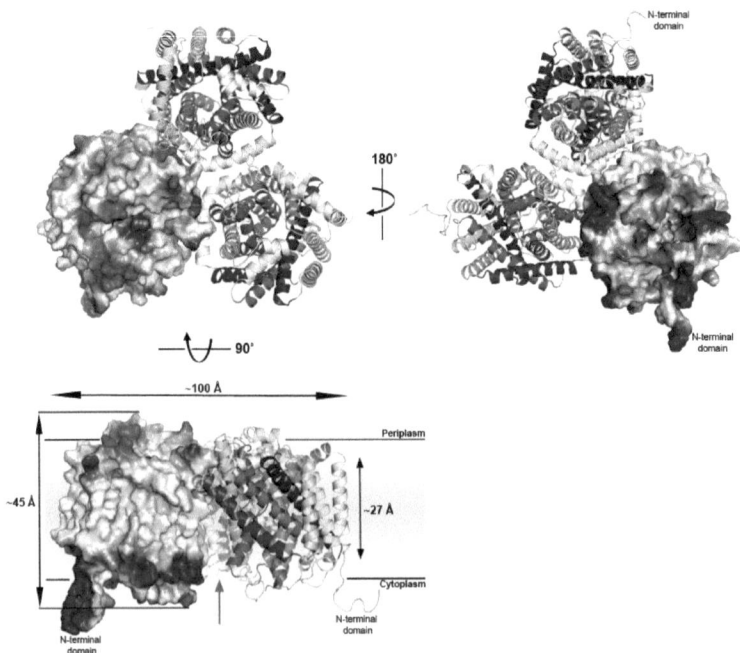

Figure 4-2 | PmCaiT trimer with surface potential shown for one of the protomers

The symmetrical CaiT trimer is shown from the periplasmic side (**A**), the cytoplasmic side (**B**) and parallel to the membrane (**C**). One protomer is shown with surface potential that was calculated at pH 6.8 with the APBS (Adaptive Poison-Boltzmann Solver), an implemented tool in PyMOL (DeLano, 2002). The longest dimension of the trimer measures ~100 Å, the distance between the periplasmatic surface and the cytoplasmatic surface is ~45 Å and the hydrophobic part of the protein measures about 27 Å in height. On the cytoplasmic side between the three protomers a hydrophobic cavity is formed (red arrow) with a height of ~24 Å and a diameter of about 10 – 15 Å, which is likely to contain lipids in the native membrane environment.

4.3.2 The CaiT protomer

The CaiT protomer consists of twelve transmembrane helices (TM1 to TM12) (Figure 4-3). Ten of these helices (TM3 – TM7 and TM8 – TM12) form two inverted repeats of five TM helices (Figure 3-24) that make up the transporter core. The helices TM1 and TM2 together with the long, curved H7 are not part of the two 5-TM inverted repeats and form the supporting scaffold for the CaiT transporter core.

Discussion

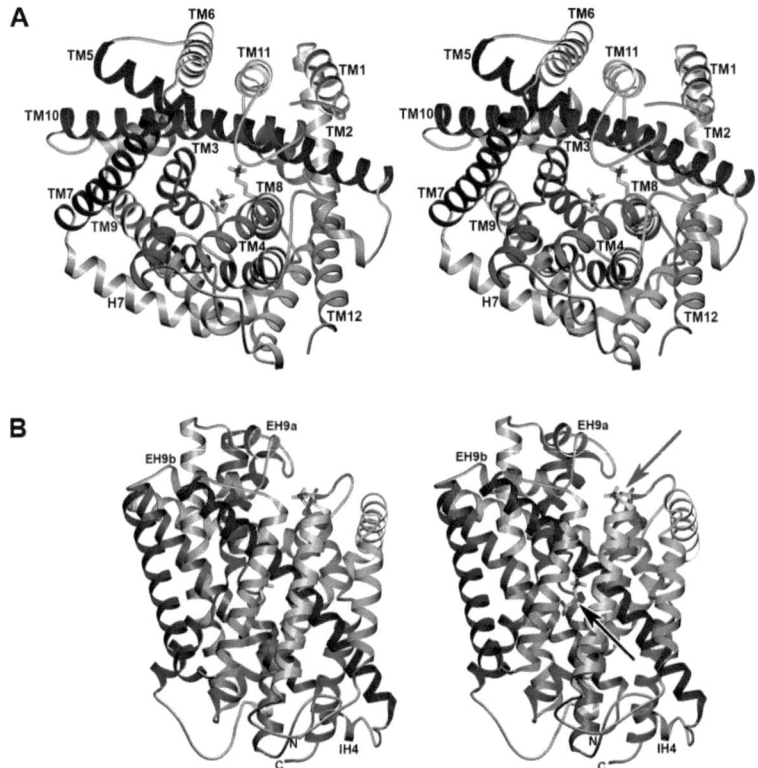

Figure 4-3 | Stereo view of the EcCaiT protomer
Periplasmic (**A**) and side (**B**) view on the EcCaiT protomer with two substrate molecules bound. The first substrate molecule is bound in the center of the protomer (black arrow), the second molecule is held in an extracellular cavity (blue arrow) near the exit of the closed substrate pathway.

The 5-TM inverted repeat motif was first reported for the sodium-coupled secondary transporter LeuT (Yamashita *et al.*, 2005). Including CaiT, there are now seven unique structures of five different transporter families all of which contain this core motif (LeuT, Neurotransmitter Sodium Symporter NSS family (Singh *et al.*, 2008; Yamashita *et al.*, 2005); vSGLT, Sodium Solute Symporter SSS family (Faham *et al.*, 2008); Mhp1, Nucleobase Cation Symporter-1 NCS-1 family (Shimamura *et al.*, 2010; Weyand *et al.*, 2008); BetP/CaiT Betaine/Carnitine/Choline Transporter

Discussion

BCCT family (Ressl *et al.*, 2009; Tang *et al.*, 2010); AdiC/ApcT, Amino acid, Polyamine and Organocation APC family (Fang *et al.*, 2009; Gao *et al.*, 2009; Gao *et al.*, 2010). The inner core of these transporters consists of an antiparallel four-helix bundle, which is formed by the first two helices of the two inverted repeats. In CaiT, the four-helix bundle provides residues for the substrate-binding site that is located in the center of the protein (Figure 3-24, Figure 3-28, Figure 4-3).

In CaiT, the two substrate-binding sites both occur on the four-helix bundle. The central transport site is located in the centre of the protein halfway across the membrane and consists of a tryptophan box (Figure 3-29, Figure 3-30), resembling that of BetP (Ressl *et al.*, 2009). A second substrate-binding site is located in an extracellular cavity near the exit of the closed substrate pathway (Figure 3-28, Figure 3-31). In the antiporter CaiT the two substrate-binding sites found in the EcCaiT structure (Figure 3-28, Figure 3-29, Figure 3-31) and the sigmoidal substrate binding behaviour (Figure 3-50) suggest that the two sites are necessary to perform the exchange of the substrate L-carnitine against the metabolic product γ-btyrobetaine. Both binding sites must be occupied, as seen in the EcCaiT structure, before substrate transport is possible.

The chemical environment of the external substrate-binding site in CaiT is conserved in BetP, suggesting a similar site could also be present in BetP. An external second substrate-binding site at a similar position was also proposed for LeuT ((Shi *et al.*, 2008); Section 4.4.2.2). In symporters like BetP and LeuT an external substrate-binding site might be necessary to deliver substrate to the transporter by binding first to its external surface. Binding of substrate to a proposed external binding site in LeuT might trigger the intracellular release of Na^+ from the Na1 binding site and substrate bound in the central substrate-binding site (Quick *et al.*, 2009; Shi *et al.*, 2008).

Discussion

4.3.2.1 Comparison of a recently published *E. coli* CaiT structure

Recently, an X-ray structure of CaiT from *E. coli* (PDB accession code 3HFX) at a resolution of 3.15 Å has been reported (Tang *et al.*, 2010). It is largely identical to the EcCaiT structure presented in this work (Figure 4-4), except from small changes in loop regions that are involved in crystal contacts.

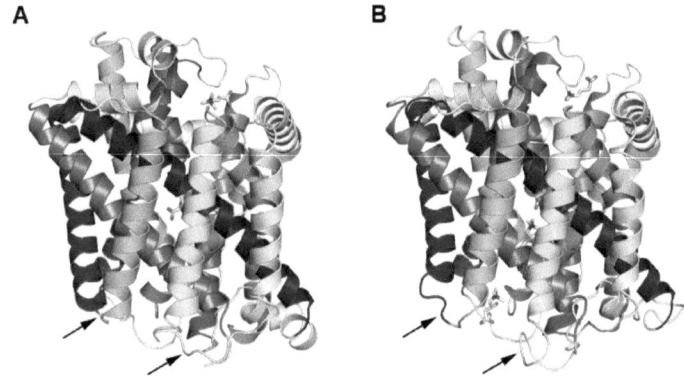

Figure 4-4 | EcCaiT protomers
The EcCaiT (**A**) structure and the *E. coli* CaiT from Tang *et al.* (**B**) are largely identical, except from changes in loop regions that are involved in crystal contacts (black arrows).

Tang *et al.* co-crystallized CaiT with 5 mM L-carnitine, which is approximately 2.5 times K_D. Each CaiT protomer binds four L-carnitine molecules, two of the binding sites confirm those that bind γ-butyrobetaine in the EcCaiT structure (Figure 4-5) presented in this work. The other two binding sites are most likely transiently occupied when the substrate moves from the cytoplasm into the central transport site.

Discussion

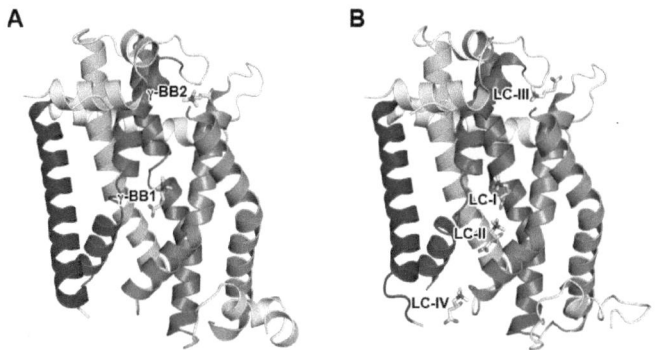

Figure 4-5 | Substrate molecules in the substrate pathway of EcCaiT

The EcCaiT substrate pathway (**A**) contains two γ-butyrobetaine molecules. The first molecule (γ-BB1) is bound in the central transport site while the second molecule (γ-BB2) is held in the regulatory binding site of the protomer. In the CaiT structure of Tang *et al.* (**B**), four L-carnitine molecules were found. The first molecule (LC-I) is localized in the central transport site, the second (LC-II) is bound ~6 Å below the first substrate molecule. The third L-carnitine molecule (LC-III) is held in the extracellular substrate-binding site and the fourth molecule (LC-IV) is bound at the entry of the cytoplasmically open substrate pathway.

The residues in the central transport site of the two CaiT structures are almost identical (Figure 4-6). Both the γ-butyrobetaine in EcCaiT and the first L-carnitine in the CaiT structure of Tang *et al.* are strongly interacting with Trp323. In both *E. coli* CaiT structures, Trp323 takes on the same rotamer orientation that allows substrate binding. This rotamer orientation is different to the one seen for Trp323 in PmCaiT at higher resolution, which has no substrate bound (Figure 3-30). The second L-carnitine froms a cation-π interaction with Tyr327 (Tang *et al.*, 2010). Tyr327 rotates by only ~2° upon L-carnitne binding, suggesting that in the inward-facing open conformation most of the residues in the pathway involved in substrate interaction have a default orientation that requires minimal rotational energy in order to transfer substrate to the central binding site.

In the external substrate-binding site (Figure 4-6), the substrate is bound by cation-π interaction to Tyr114 and Trp316 in both CaiT structures. In EcCaiT, γ-butyrobetaine forms an additional hydrogen bond to the backbone amid nitrogen of

Gly315 while in the other *E. coli* CaiT structure an additional hydrogen bond is formed by L-carnitine to Gly310 (Tang *et al.*, 2010).

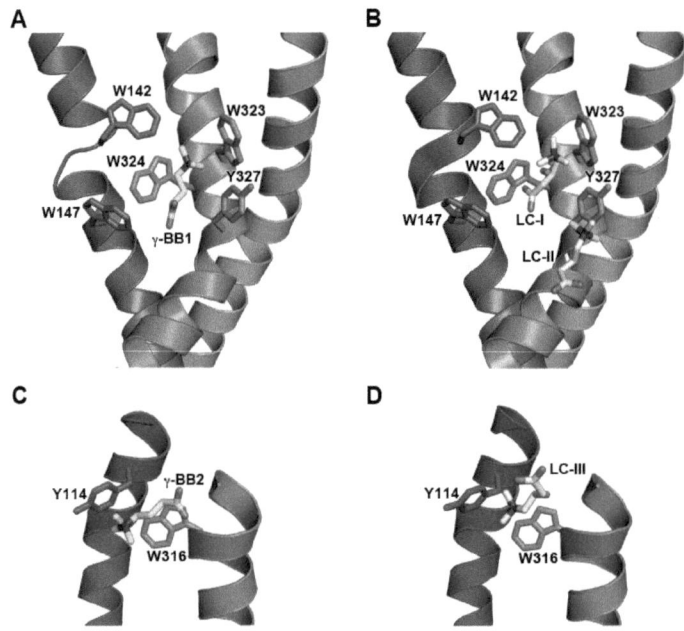

Figure 4-6 | Central transport site and regulatory binding site

The residues in the central transport site of EcCaiT (**A**) and in the *E. coli* CaiT reported by Tang *et al.* (**B**) are almost identical. The substrate γ-butyrobetaine (γ-BB1) or L-carnitine (LC-I) interacts mainly by cation-π interaction with Trp323, which takes on the same rotamer orientation in both structures. The second L-carnitine molecule (LC-II) bound in CaiT (Tang *et al.*, 2010) interacts with Tyr327 that only reorients by ~2° upon substrate binding. In the regulatory substrate-binding site of EcCaiT (**C**) the γ-butyrobetaine interacts by cation-π interaction with Tyr114 and Trp316. The same residues interact with L-carnitine (LC-III) in the external substrate-binding site of CaiT reported by Tang *et al.* (**D**).

In both *E. coli* CaiT structures, the substrates L-carnitne and γ-butyrobetaine are bound in the same way mainly by cation-π interactions in both substrate-binding sites. This is the reason why both substrates can be transported in both directions. Additional hydrogen bonds between the carboxyl group of the substrate and residues

Discussion

in the protein stabilize the orientation of the substrate in the binding site. The formation of specific hydrogen bonds is dependent on the substrate flexibility.

4.3.3 Transporter conformation

In the alternating access mechanism the central transport site must be accessible from one side of the membrane to bind substrate. To transport the substrate across the membrane, the protein must undergo conformational changes in which the transporter closes the central transport site from the one site of the membrane before it subsequently opens to the other site allowing substrate transport without opening a continuous transmembrane pore (Jardetzky, 1966; Krishnamurthy et al., 2009). The four conformations (Figure 1-10) that were previously described are the outward-facing open, represented by Mhp1 (PDB accession code 2JLN; (Weyand et al., 2008)), the outward-facing occluded, represented by LeuT-Leu (PDB accession code 2A65; (Yamashita et al., 2005)), the occluded, represented by BetP-Bet (PDB accession code 2WIT; (Ressl et al., 2009)) and the inward-facing occluded, represented by vSGLT-Gal (PDB accession code 3DH4; (Faham et al., 2008)). CaiT crystallized in an inward-facing, open conformation. Access to the central transport site from the cytoplasm is unhindered by sidechains. The inward-facing, open conformation was the last major conformation missing in the alternating access transport mechanism of LeuT-type transporters. Superpositions of the different transporters with PmCaiT or EcCaiT show the major conformational changes in the helices lining the substrate transport pathway (TM3, TM7, TM8, TM10 and TM12, CaiT numbering) of the transporters. The root means square deviation (rmsd) values of the different transporter conformations are listed in Table 4-1.

Discussion

Table 4-1 | Structure comparisons of PmCaiT and EcCaiT to each other and to the structurally related transporters BetP, vSGLT, LeuT and Mhp1

Mol A	Mol B	Z-Score	Aligned Residues	rmsd (Å)	Sequence Identity (%)
PmCaiT (Trimer)	EcCaiT (Trimer)	53.6	972	0.9	87
PmCaiT (Protomer)	EcCaiT (Protomer)	60.5	496	0.7	87
PmCaiT (Trimer)	BetP (Trimer)	43.1	1348	2.2	25
PmCaiT (Monomer)	BetP (Monomer)	43.7	478	2.2	25
EcCaiT (Trimer)	BetP (Trimer)	37.2	895	2.3	25
EcCaiT (Monomer)	BetP (Monomer)	44.0	481	2.2	25
PmCaiT (Monomer)	vSGLT (Monomer)	13.7	334	4.2	13
EcCaiT (Monomer)	vSGLT (Monomer)	13.6	333	4.0	12
PmCaiT (Monomer)	LeuT (Monomer)	14.8	342	4.1	10
EcCaiT (Monomer)	LeuT (Monomer)	14.7	335	4.3	10
PmCaiT (Monomer)	Mhp1 (Monomer)	17.2	320	4.2	8
EcCaiT (Monomer)	Mhp1 (Monomer)	16.8	317	4.4	7

Discussion

In the different transporters the two inverted repeats start with different helices. In CaiT and BetP, the first two TM helices (TM1 and TM2) form a protecting scaffold for the protein core and the inverted repeats start with TM3. In LeuT, the supporting helices are located at the C-terminal end (TM11 and TM12) and the inverted repeats start with TM1. vSGLT has 14 TM helices in which the first (TM1) and the last three helices (TM12-TM14) form the protein scaffold and the inverted repeats start with TM2. Table 4-2 provides a list of residues that correspond to the 10 TM helices of the two inverted repeats.

Table 4-2 | Lookup table of residues for the 10 TM helices of the two inverted repeats in LeuT-type transporters

	CaiT	BetP	vSGLT	LeuT	Mhp1	AdiC	ApcT
	Residues	Residues	Residues	Residues	Residues	Residues	Residues
Repeat 1							
Helix 1	87 – 118	137 – 169	52 – 80	10 – 38	28 – 55	11 – 37	9 – 37
	TM3	TM3	TM2	TM1	TM1	TM1	TM1
Helix 2	127 – 163	177 – 212	83 – 109	40 – 72	57 – 86	43 – 67	40 – 66
	TM4	TM4	TM3	TM2	TM2	TM2	TM2
Helix 3	186 – 224	234 – 268	123 – 158	87 – 125	99 – 137	81 – 112	83 – 117
	TM5	TM5	TM4	TM3	TM3	TM3	TM3
Helix 4	228 – 249	275 – 296	161 – 178	165 – 185	142 – 159	122 – 143	122 – 141
	TM6	TM6	TM5	TM4	TM4	TM4	TM4
Helix 5	251 – 277	300 – 325	185 – 213	189 – 214	161 – 191	146 – 172	145 – 172
	TM7	TM7	TM6	TM5	TM5	TM5	TM5
Repeat 2							
Helix 6	311 – 340	358 – 390	249 – 277	240 – 269	208 – 234	195 – 217	183 – 214
	TM8	TM8	TM7	TM6	TM6	TM6	TM6
Helix 7	343 – 377	393 – 427	279 – 314	275 – 306	241 – 278	226 – 248	218 – 247
	TM9	TM9	TM8	TM7	TM7	TM7	TM7
Helix 8	403 – 435	448 – 482	348 – 385	336 – 371	295 – 331	277 – 310	269 – 305
	TM10	TM10	TM9	TM8	TM8	TM8	TM8
Helix 9	445 – 467	488 – 511	391 – 418	374 – 396	335 – 351	323 – 342	320 – 337
	TM11	TM11	TM10	TM9	TM9	TM9	TM9
Helix 10	469 – 502	513 – 546	422 – 448	398 – 425	359 – 383	351 – 376	339 – 364
	TM12	TM12	TM11	TM10	TM10	TM10	TM10

4.3.3.1 Alternating access models

For the alternating access mechanism of LeuT-type transporters two models have been proposed. Gouaux and coworkers (Krishnamurthy *et al.*, 2009; Singh *et al.*, 2008) propose a mechanism for LeuT based on the outward-facing open conformation of LeuT-Trp (Singh *et al.*, 2008), the outward facing occluded conformation of LeuT-LeuT (Yamashita *et al.*, 2005) and the inward facing occluded conformation of vSGLT (Faham *et al.*, 2008) in which the unwound regions of TM1 and TM6 provide a hinge around which rotational and translational conformational changes can occur. The continuous helices TM2 and TM7 would constrain large movements of TM1 and TM6 but would move and bend in the transition from the outward-facing to the inward-facing conformation. In addition, a bending of TM3 and TM8 may also contribute to the opening and closing of the transporter (Figure 4-7 black circles; (Krishnamurthy *et al.*, 2009)). The helices which form a scaffold around the four-helix bundle stabilizes the transporter within the lipid membrane and mediate conformational changes of the transporter on one site of the membrane to movements on the other side (Krishnamurthy *et al.*, 2009).

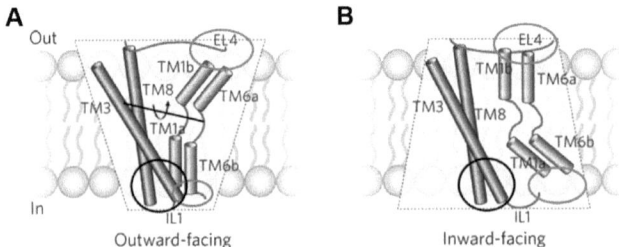

Figure 4-7 | Proposed mechanism of the transition from the outward-facing to the inward-facing conformation in LeuT-type transporters

The helices TM1, TM3, TM6 and TM8 line the central transport pathway. The extracellular and intracellular loops EL4 and IL1, respectively, act as lids that seal the extracellular and intracellular pathway in their closed conformation. In the transition from the outward-facing conformation (LeuT, **A**) to the inward-facing conformation (vSGLT, **B**) TM1 and TM6 rotate relative to TM3 and TM8. The rotation axis is shown in **A** in black. The figure is adapted from (Krishnamurthy *et al.*, 2009).

Discussion

In this proposed mechanism, the helices of the four-helix bundle would move independently from one another and not as a rigid body (Krishnamurthy *et al.*, 2009) as it is proposed in a second transport mechanism.

The second mechanism is described as rocking bundle mechanism, which is based on the LeuT structure, LeuT modeled in the inward-facing conformation as well as mutagenisis and accessibility studies of the mammalian serotonin transporter SERT, which is a member of the same transporter family as LeuT (Forrest and Rudnick, 2009; Forrest *et al.*, 2008). The four-helix bundle of LeuT (Table 4-2) is seen as a rigid body (Figure 4-8) and the transporter transforms from the outward-facing to the inward-facing conformation by a rotation of the four-helix bundle relative to the scaffold of the protein around an axis that is close to the substrate and ion binding site (Forrest *et al.*, 2008). During the substrate transport, the rotation of the four-helix bundle can be regarded as an alternate rocking of the bundle (Forrest *et al.*, 2008) from one conformation to the other.

Discussion

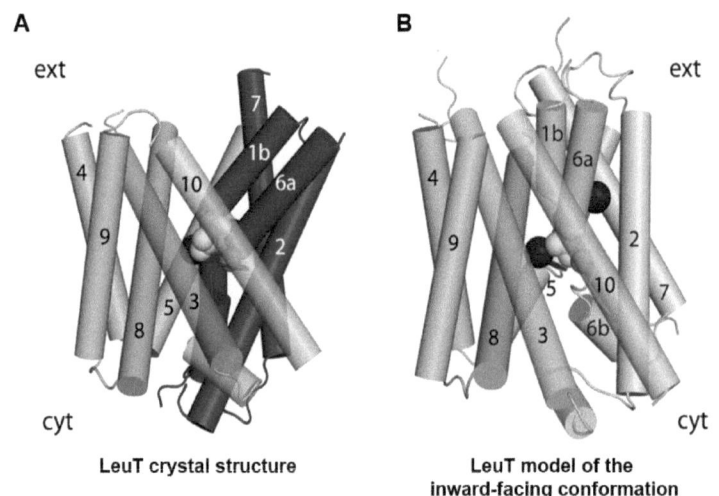

Figure 4-8 | Conformational changes of LeuT predicted by the rocking bundle model

LeuT is shown in the outward-facing conformation (**A**) in which it was crystallized and in the modelled inward-facing conformation (**B**). The scaffold of the protein is shown in blue, and the four-helix bundle in red for the outward-facing conformation and in gold for the inward-facing conformation. In the transition from the outward-facing conformation to the inward-facing conformation, the four-helix bundle rotates around an axis that lies perpendicular to the paper plane and close to the substrate and ion binding site. The substrates leucine (yellow and dark blue) and sodium (green) are shown in space-filling representation. The figure is adapted from (Forrest *et al.*, 2008).

Recently, Mhp1 has been crystallized in three different conformations, the outward-facing open, the outward-facing occluded and the inward-facing open conformation (Shimamura *et al.*, 2010; Weyand *et al.*, 2008). Based on the analysis of the three structures and MD simulations with the protein in a nativelike bilayer the authors propose a mechanism similar to the rocking bundle mechanism in which the scaffold and the four-helix bundle move relative to each other as approximately rigid bodies (Shimamura *et al.*, 2010). However, two of the three Mhp1 structures, the outward-facing occluded conformation (4.0 Å) and the inward-facing open conformation (3.8 Å), were solved at low resolution at which precise helix positioning and detailed analysis of helix movements are not possible.

Discussion

The rocking bundle model combines the LeuT-type transporter architecture with the internal symmetry of the two 5-TM inverted repeats and the arrangement of the four-helix bundle relative to the transporter scaffold with symmetric movements resulting in conformational changes that expose the central substrate-binging site alternately to the two sites of the membrane. By definition, models describe complex mechanisms in a very simplified manner.

Comparisons of the two CaiT structures and BetP show that the transition from the inward-facing open conformation to the occluded conformation involves iris-like movements of the helices TM3 and TM8 in the four-helix bundle as well as TM7 and TM10, which are located in the scaffold. These helix rotations are similar to the movements proposed by Gouaux and coworkers (Krishnamurthy *et al.*, 2009; Singh *et al.*, 2008). The transition from the occluded conformation to the outward-facing open conformation, represented by LeuT-Trp (Singh *et al.*, 2008), requires an additional independent shift of the periplasmatic part of TM12. These movements are inconsistent with the rocking bundle model.

The substrate transport mechanism seems to vary among the different LeuT-type transporters and can only be described in detail for a single transporter when the structures of the three major conformations are available at high resolution.

Discussion

4.4 The CaiT transport mechanism

4.4.1 Cation-π interaction

In CaiT the specific binding of the substrates appears to be mainly established by cation-π interactions between the positively charged quaternary ammonium group of the substrate and aromatic residues like tryptophans or tyrosines of CaiT (Figure 4-11, Figure 4-12).

The origin of the cation-π interaction is the negative electrostatic potential of the quadrupole moment in sp^2 carbon systems such as benzene or tryptophan. This negative electrostatic potential is highest in the center of the aromatic ring (Figure 4-9) and arises from the separation of the six local $C^{\delta -}$-$H^{\delta +}$ bond dipoles (Dougherty, 2007). Cations are attracted to the negative electrostatic potential over the face and not the edge, of the aromatic system (Dougherty, 1996).

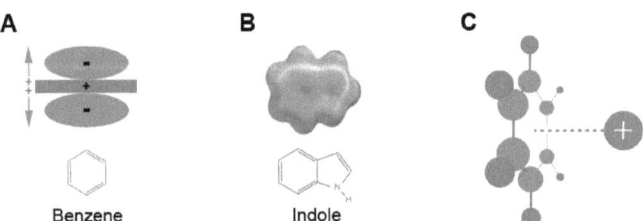

Figure 4-9 | Cation-π interaction

The cation-π interaction is due to an electrostatic attraction between a positive charge and the quadrupole moment of an aromatic system. The quadrupole moment of the three sp^2 orbitals of benzene, viewed on edge, with regions of positive (blue) and negative (red) partial charge, is shown schematically in (**A**). The quadrupole moment arises from the separation of the six local $C^{\delta -}$-$H^{\delta +}$ bond dipoles. High-level quantum mechanical calculations (**B**) show the negative electrostatic potential (red) in the center of the trytophan indole ring to which the cations bind. Cations are attracted to the negative electrostatic potential over the face, not the edge, of the indole ring (**C**). (Dougherty, 1996, 2007)

Discussion

One important role of the cation-π interaction is stabilizing the secondary structure of proteins by establishing a specific binding force between the amino acids Phe/Typ/Trp and Lys/Arg. Another role of the cation-π interaction is protein-ligand recognition. Strong cation-π interactions often involve ligands that contain a quaternary ammonium ion, such as $RNMe_3^+$ (Dougherty, 2007). The orientation of the ligand relative to the aromatic ring influences the binding strength. The strongest cation-π interaction involves a T-shape geometry of the ligand relative to the aromatic system (Figure 4-10; (Gallivan and Dougherty, 1999)).

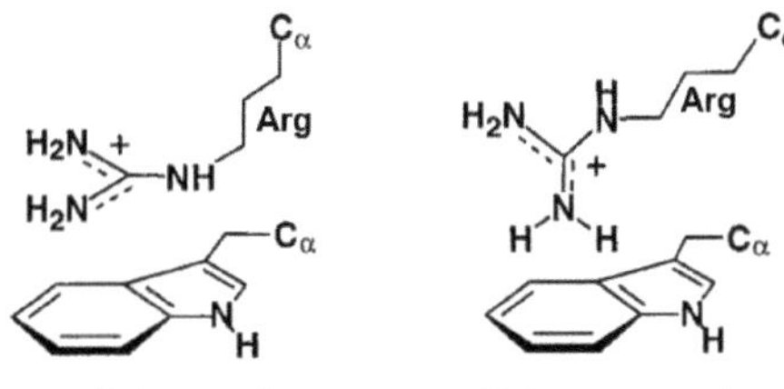

Figure 4-10 | Geometry of cation-π interactions

The preferred geometry of cation-π interactions in protein structures is the orientation of the cation parallel to the quadrupole moment of the tryptophan indole ring, although some of the strongest cation-π interactions involve the T-shape geometries (Gallivan and Dougherty, 1999).

4.4.1.1 Cation-π interaction in CaiT

In EcCaiT, the γ-butyrobetaine bound in the central transport site is oriented in the T-shape geometry to Trp323 (Figure 4-11), forming a strong cation-π interaction with the tryptophan. In PmCaiT, Trp323 is rotated by ~60° to an orientation that prohibits a cation-π interaction to the substrate but allows hydrogen bond formation to small solvent molecules. The indole ring of Trp323 only reorients into the substrate-binding position when a substrate molecule binds in the external binding site. The positively charged quaternary ammonium group of the substrate replaces the water molecules in the central transport site and interacts strongly with

Discussion

the indole ring of Trp323 since the sp^2 carbon is more electronegative than hydrogen (Dougherty, 2007).

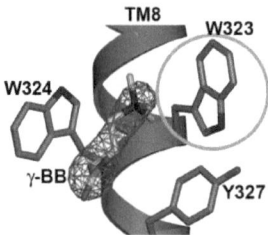

Figure 4-11 | Geometry of the cation-π interaction in the central transport site of EcCaiT

The positively charged quaternary ammonium group of γ-butyrobetaine in the central transport site of EcCaiT is oriented in a T-shape geometry to Trp323, which allows a strong cation-π interaction.

In the external regulatory binding site of EcCaiT, the γ-butyrobetaine is bound in parallel geometry to Tyr114 and W316 (Figure 4-12). The parallel cation-π interaction is weaker than the interaction in the T-shape geometry (Gallivan and Dougherty, 1999).

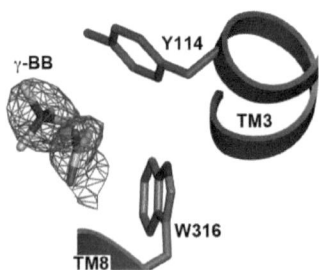

Figure 4-12 | Geometry of the cation-π interaction in the regulatory binding site of EcCaiT

The cation-π interaction of the positively charged quaternary ammonium group of γ-butyrobetaine to Tyr114 and Trp316 in the regulatory binding site of EcCaiT is in parallel geometry.

The chemical environment and the binding characteristics of the substrates in the two substrate-binding sites are different. The binding environment in the central

Discussion

transport site is provided by four tryptophans and one polar interaction to Met331. In the regulatory binding site, two aromatic residues provide the binding environment. An additional hydrogen bond is also formed between the carboxyl group of the substrate and the backbone amide nitrogen of G315 to hold the substrate in position in the external binding site.

Based on the chemical environment and the binding characteristic of the substrate in the CaiT structure, the central transport site should have a significantly higher affinity for the substrate than the external binding site. However, in CaiT substrate binding is cooperative (Sections 3.6.2.2 and 4.4.2.2) and the substrate binding affinity in the central transport site is probably determined by the substrate affinity of the external binding site. The positive cooperative binding of substrate in CaiT (Figure 3-50) shows a heterotropic effect, i.e., after binding of one substrate to the external site the substrate binding affinity of the central binding site is strongly increased and as long as substrate is bound in the external binding site, the central transport site has a strong affinity for the substrate. The increased substrate-binding affinity of the central binding site after binding of substrate in the external binding site can be explained by small conformational changes in the protein – a mechanism known as Koshland's induced fit model (Koshland, 1958, 1973; Koshland *et al.*, 1966). According to this mechanism, the protein is assumed to exist in two conformations. In a heterotropic positive cooperative induced fit mechanism the absence of substrate causes the protein to stay in an inactive conformation. However, when substrate is available it binds to the regulatory binding site and induces a conformational change which favours binding of a second substrate at a second binding site (Koshland, 1973).

The positive cooperative substrate binding suggests an allosterically regulated transport mechanism of CaiT, which is likely to be the reason why the calculated K_D in the binding studies (Table 3-4, Table 3-5, Table 3-6) and the determined K_M in the kinetic analyses (Table 3-2, Table 3-3) show a difference of a factor of ~50.

4.4.2 Na$^+$-independent and cooperative substrate/product antiport in CaiT

The structure of CaiT gives the first example of a secondary transporter that is independent of an electrochemical gradient. All other structurally related LeuT-type transporters are either Na$^+$- or H$^+$-dependent. Structural analysis combined with binding and transport studies reveal how the antiporter CaiT transports its substrates without the help of a cation.

4.4.2.1 Sodium-independent substrate transport mechanism

In the high-resolution X-ray structure of LeuT, two well-resolved sodium binding sites (Na1 and Na2) have been determined (Yamashita *et al.*, 2005). The two sodium ions have key roles in binding the substrate and in stabilizing the architecture of the central transport site. None of the three CaiT structures shows density for a Na$^+$ ion in the vicinity of either of the two corresponding Na sites of LeuT. Instead, inspection of side chain residues in these two regions provides an explanation for the Na$^+$-independence of CaiT.

The binding of γ-butyrobetaine in the central binding site of EcCaiT reveals an attractive interaction between the carboxyl group of the γ-butyrobetaine, and the sulfur of Met331 (Figure 4-11). Although the methionine side chain is usually thought of as hydrophobic, it can in fact participate in polar interactions because the large uncharged sulfur is more easily polarized than smaller atoms. Two different types of interactions with neighbouring charges are possible (Rosenfield *et al.*, 1977). The methionine sulfur is either negatively polarized and behaves as a nucleophile towards positively charged binding partners (e.g. Na$^+$, Ni^{2+}, H$^+$, C$^+$). Alternatively, the methionine sulfur can be positively polarized and behave as an electrophile towards anions or atoms with a partial negative charge (e.g. Cl$^-$, O$^-$). In CaiT, the sulfur of Met331 interacts specifically with the carboxyl group of γ-butyrobetaine.

Discussion

This provides an elegant solution to the problem of recognizing and binding a hydrophilic compound in the hydrophobic interior of a protein without the need for an ion as a second non-protein binding partner. It is known from the structures of small organic compounds like 3-(methylthio) propanoic acid or norbornane endo-acid (Mahling *et al.*, 1987) that an sulfur atom in a covalent bond, as in methionine, can interact with carboxylates (Burling and Goldstein, 1993; Mahling *et al.*, 1987; Pal and Chakrabarti, 2001).

Met331 is conserved in the prokaryotic CaiT and in the mammalian organic cation/carnitine transporters (OCTN) (Figure 6-2). In BetP, however, the position of the methionine is taken by Val381, which cannot coordinate the negatively charged carboxyl group of the substrate. Mutating Met331 to valine in PmCaiT reduces K_D for substrate binding by a factor of ~4, but v_{max} in transport measurements by a factor of 10 (Figure 3-49, Table 3-5, Figure 3-41, Table 3-3). This highlights the pivotal role of this residue for transport. However, the mutation does not make CaiT Na^+-dependent.

Interestingly, ApcT, which is H^+-dependent, but, like CaiT Na^+-independent, also has a methionine (Met202) in the Na1 position (Figure 4-13; (Shaffer *et al.*, 2009). The putative central transport site of ApcT was not occupied by substrate so that an interaction between this methionine and the substrate was not observed. Instead, the volume of the putative substrate-binding site was occupied by ordered water molecules (Figure 4-13; (Shaffer *et al.*, 2009)).

Discussion

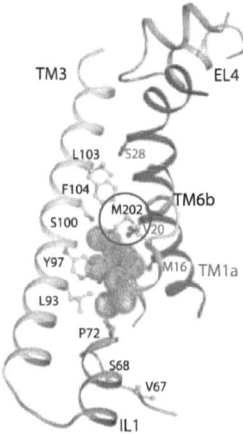

Figure 4-13 | Putative central transport site in ApcT

In the putative central transport site of the H⁺-dependent Apc transporter, Met202 (red circle) might coordinate the substrate as in CaiT. In this way, the transporter would avoid the need for an additional binding partner. This picture is adapted from (Shaffer *et al.*, 2009).

The Na2 position in LeuT is occupied by Arg262 (TM7) in CaiT (Figure 4-14). This residue is accessible to solvent in the open cytoplasmic funnel. The positively charged arginine sidechain evidently has the same role as Na2 in LeuT, linking TM3 to TM10 and stabilizing the unwound region of TM3, which contributes to substrate binding in the external binding pocket through Tyr114 (Figure 4-12). In the H⁺-dependent ApcT, a lysine (K158) is located in a position equivalent to the Na2 ion in LeuT (Figure 4-14; (Shaffer *et al.*, 2009)). This residue undergoes reversible protonation and deprotonation during the substrate transport cycle (Shaffer *et al.*, 2009).

Replacing Arg262 by glutamate in PmCaiT renders the protein inactive but with addition of Na⁺ the mutant regains a partial transport activity that is four-fold reduced compared to wildtype PmCaiT (Figure 3-43, Table 3-3). The insertion of a negative charge at a position that was held by a positive charge in the wildtype protein has to make the protein inactive, as the chemical environment is adjusted to

Discussion

compensate a positive charge. However, it is striking that the inactive PmCaiT Arg262Glu mutant can be partially rescued by addition of Na^+. Thus, a single point mutation confers Na^+-dependence in CaiT. However, it is unlikely that the mutant cotransports Na^+ with the substrate. It is more likely that the Na^+ remains bound the to Glu262 in the mutant. Detailed Na^+ stoichiometry analyses of the Na^+-dependent PmCaiT Arg262Glu mutant and control experiments with a non-charged residue introduced at this position still need to be done. These experiments will clarify if a positive charge is essential or if a neutral environment, with only one Na^+ compensating for the negative charge of the glutamate, is sufficient for protein activity.

Discussion

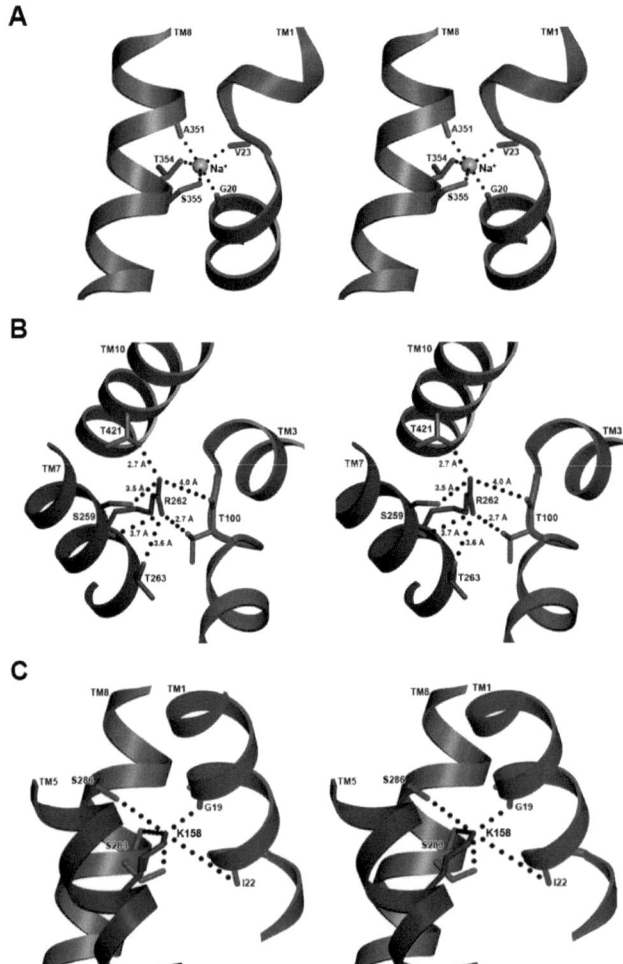

Figure 4-14 | Position of Na2 in LeuT and the equivalent position in the Na$^+$-independent transporters CaiT and ApcT

Na2 (green sphere) in LeuT (**A**) links TM1 to TM8. These helices correspond to helices TM3 and TM10 in CaiT. The sodium ion stabilizes the unwound region of TM1. TM1 in LeuT provides residues for substrate binding in the central transport site. The position equivalent to Na2 in LeuT is taken by Arg262 in CaiT (**B**) and Lys158 in ApcT (**C**). In both Na$^+$-independent transporters, a positively charged side chain takes over the function of Na$^+$. The protonated side chain of Arg262 avoids the need of an additional binding partner in CaiT, whereas the Lys158 in the H$^+$-dependent ApcT needs to undergo reversible protonation and deprotonation during the transport cycle.

Discussion

4.4.2.2 Cooperative activation of CaiT

The EcCaiT crystal structure revealed two substrate-binding sites. Substrate-binding curves for CaiT reconstituted into proteoliposomes are sigmoidal (Figure 3-50), indicating positive cooperativity (Ricard and Cornish-Bowden, 1987). This means that transport activity is allosterically regulated, such that a substrate molecule bound to one site increases the binding affinity of the other. The cooperativity is not apparent with detergent-solubilized CaiT (Figure 3-47, Figure 3-48, Figure 3-49), presumably because the transporter is more relaxed and flexible in detergent solution than in the membrane. Delipidation upon detergent solubilization during protein preparation decreases the lateral pressure that impacts on the protein in the lipid bilayer (Palsdottir and Hunte, 2004). Thus a lipid environment that may include ordered lipid in the hydrophobic cavity of the trimer (Figure 3-25, Figure 3-26, Figure 4-2) seems to be required for cooperativity.

It is reasonable to assume that the external substrate-binding site is the regulatory site. This assumption is strongly supported by the finding that the PmCaiT structure does not contain bound substrate, whereas that of EcCaiT does. In PmCaiT, access to the second site is blocked by a crystal contact (Figure 3-32), which cannot form when substrate is bound. An empty regulatory site would reduce the affinity of the central transport site, which indeed is also empty in PmCaiT.

The affinity of CaiT for γ-butyrobetaine is significantly lower than for L-carnitine (Table 3-6) and this must reflect the binding properties of the regulatory site. With its extra hydroxyl group, a L-carnitine molecule can establish another hydrogen bond to the ε-amino group of Lys470 in the regulatory binding site (Figure 4-12), whereas there are no such polar interactions in the central transport site. The finding that the W316 mutant is strongly impaired in transport but only moderately in substrate binding (Table 3-3, Table 3-5, Table 3-6) supports the conclusion that the external binding site is the regulatory site, as the structure of the transport site, and therefore its binding affinity for either substrate, will be largely unaffected by the

Discussion

W316A mutation. If the external site were not the regulatory site, no strong effect on transport would be expected, contrary to what is observed (Figure 3-49, Figure 3-50).

The Hill coefficients derived from the binding curves (Figure 3-50) range from 1.4 or 1.5 for wt PmCaiT or EcCaiT and L-carnitine, to 1.7 for the W316 mutant and γ-butyrobetaine (Table 3-6), indicative of pronounced cooperativity. Many allosteric proteins are oligomers with only one binding site per protomer. CaiT has two binding sites and it is therefore interesting to ask whether the cooperativity applies to the two binding sites in the protomer, or to all six in the trimer. Comparison of the EcCaiT and PmCaiT trimers shows that in addition to structural changes in the protomer each protomer tilts 3° toward the three-fold axis of the trimer (Figure 4-15) when substrate is bound.

The overall rms deviation between the substrate-bound and the apo protomer is 0.7 Å and 0.9 Å for the trimers (Table 4-1) with and without bound substrate. This difference indicates that these changes are significant as the rmsd between spinach and pea LHC-II, two similarly homologous, trimeric membrane proteins that do not undergo a conformational change, is 0.35Å (Barros *et al.*, 2009). A Hill coefficient >1 signifies an interaction of at least 2 binding sites, which are likely to be those observed in the protomer, as cooperativity between all six sites in the trimer should result in a higher Hill coefficient. However, the exact role of the CaiT trimer in allosteric regulation remains to be determined.

Discussion

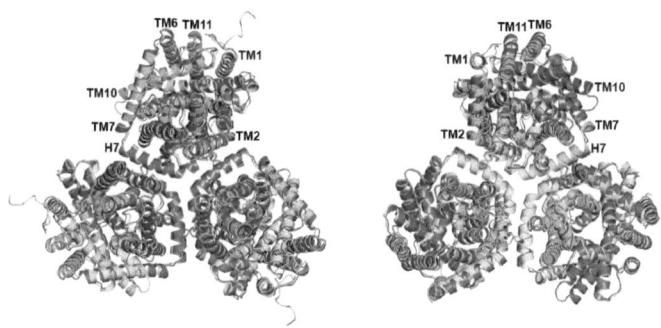

Figure 4-15 | Superposition of the PmCaiT trimer on the EcCaiT trimer
Cytoplasmice (left) and extracellular (right) view of the PmCaiT trimer (yellow) superimposed on the EcCaiT trimer (blue). The superposition indicates a 3° tilt of the substrate-bound EcCaiT protomer relative to PmCaiT towards the threefold axis.

A second substrate-binding site has also been proposed for LeuT. The location of this site in LeuT was deduced from stirred molecular dynamic (SMD) simulations (30 ns) in which the substrate leucine was pulled from the occluded position along the possible substrate pathway towards the extracellular region. This site is located in an extracellular vestibule around 11 Å above the central transport site (Quick *et al.*, 2009; Shi *et al.*, 2008). SMD simulations of LeuT together with kinetic substrate trapping studies suggest that binding of substrate at the extracellular binding site allosterically triggers intracellular Na^+ and substrate release from the central binding site (Shi *et al.*, 2008). Three different tricyclic antidepressants (TCAs) or an *n*-octyl-β-D-glucopyranoside (OG) molecule bound at this position renders the protein inactive (Quick *et al.*, 2009; Singh *et al.*, 2008; Zhou *et al.*, 2007).

Discussion

4.4.2.3 Substrate translocation model

Analysis of the two structures, EcCaiT and PmCaiT, combined with transport and binding studies, allows to propose a mechanism of allosterically regulated substrate uptake by CaiT that encompasses all the findings reported in this thesis (Figure 4-16). This model should be regarded as speculative, until structures of all intermediate stages have been determined. According to this proposed mechanism, a substrate molecule first binds to the regulatory extracellular site. This substrate would normally be L-carnitine, which is more abundant outside the cell than its metabolic product γ-butyrobetaine. Through the network of hydrogen bonds and hydrophobic sidechain contacts, substrate binding to the regulatory site triggers the reorientation of Trp323, which would result in an increased affinity for substrate in the central transport site. The wide open cytoplasmic substrate funnel in the default low-energy state of CaiT ensures free access from the cell interior, so that substrate can bind as soon as Trp323 reorients. The substrate molecule binding from the inside would be γ-butyrobetaine, which is more abundant in the cell.

Discussion

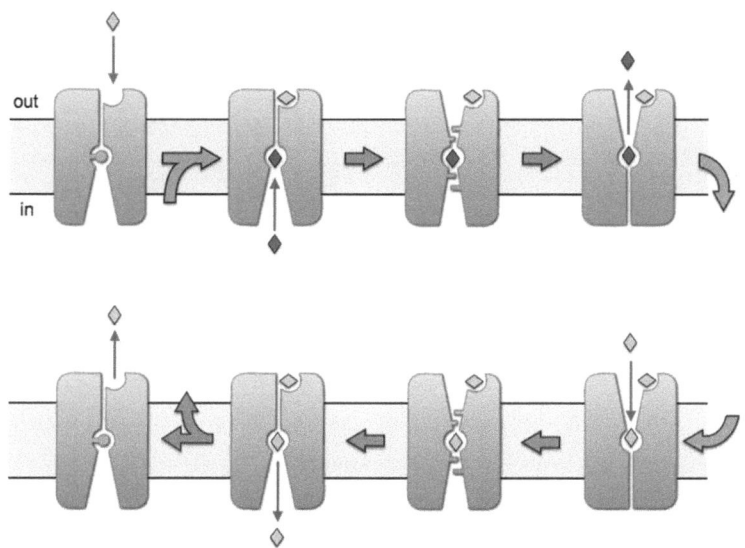

Figure 4-16 | Proposed substrate antiport mechanism
The empty transporter, represented by the structure of PmCaiT, binds a L-carnitine molecule (yellow diamond) in its regulatory site. This triggers the reorientation of Trp323, so that a γ-butyrobetaine molecule (red diamond) can bind in the central transport site. In the outside-open state, the γ-butyrobetaine diffuses out of the central binding site, and L-carnitine binds instead. The transporter then changes conformation back to the inside-open state, releasing the L-carnitine into the cell interior, where it is consumed. When the L-carnitine concentration outside the cell drops below a critical level, the substrate diffuses out of the regulatory site, and the transporter switches off.

The subsequent steps in the CaiT transport cycle can be gleaned from the structure of BetP and the LeuT-type transporters, which have different default low-energy states shown by their crystal structures. Transition from the inside-open to the occluded state, represented by BetP, involves an iris-like movement of the helices TM3 and TM8 in the central bundle and TM7 and TM10 of the protein scaffold (Figure 4-17), along with the reorientation of bulky hydrophobic sidechains that lock the substrate into the transport site.

Discussion

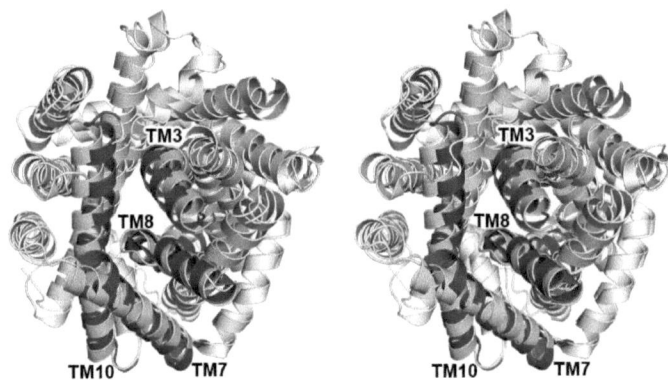

Figure 4-17 | Iris-like helix movement in the transition from the inward-facing, open conformation to the occluded conformation

Stereo drawing of superposed protomers of PmCaiT and BetP, showing an iris-like movement of the cytoplasmatic part of the helices TM3, TM7, TM8 and TM10 (yellow for PmCaiT, red for BetP) in the transition from the inward-facing, open conformation of CaiT to the occluded state of BetP.

The outside-open conformation is exemplified by LeuT and its relatives that crystallized in the same state (Faham *et al.*, 2008; Fang *et al.*, 2009; Gao *et al.*, 2009; Ressl *et al.*, 2009; Shaffer *et al.*, 2009; Shimamura *et al.*, 2010; Singh *et al.*, 2008; Tang *et al.*, 2010; Weyand *et al.*, 2008; Yamashita *et al.*, 2005). Although sequence similarity between the LeuT-type transporters is low (Table 4-1), the overall structures of the protomers are surprisingly well conserved, and a superposition of ten TM helices of the transporter core (Table 4-2) allows describing the helix movements in the transition from the occluded to the outside-open conformation. When modelled on the outside-open conformation of LeuT (Figure 4-18), the extracellular ends of helices TM3 and TM4 in repeat 1 of CaiT move by 3.2 Å and 5 Å, while TM8 and TM12 in repeat 2 move by 6.2 Å and 5 Å, in each case out of the extracellular substrate pathway to open it. It follows that the hydrogen bonds between Glu111 and Tyr320, and between Tyr137 and Thr317 (Figure 3-33) must break for the outside funnel to open. Several of the hydrogen bonds around Glu111 are longer by up to 1.0 Å in EcCaiT than in PmCaiT, suggesting that substrate binding to the external site weakens the central hydrogen bonding network in the protomer. Therefore the

Discussion

transition to the outside-open conformation would require less energy in EcCaiT, which contains bound substrate, than in PmCaiT, which does not.

In the outside-open state, γ-butyrobetaine would diffuse out of the transport site, and the more abundant L-carnitine would bind instead. The process would then go into reverse. The extracellular pathway would close, and the cytoplasmic funnel would open, releasing L-carnitine to the cell interior where its concentration is low. The constant consumption of L-carnitine in the cell would maintain a concentration gradient that drives uptake. In the CaiT structure modelled on the outward-open state of LeuT, the second site remains intact (Figure 4-18) suggesting that the substrate remains bound throughout the transport cycle (Figure 4-16) as long as there is sufficient substrate in the environment. When the external L-carnitine concentration drops below a critical level, the substrate would diffuse out of the regulatory site and transporter would switch off.

Discussion

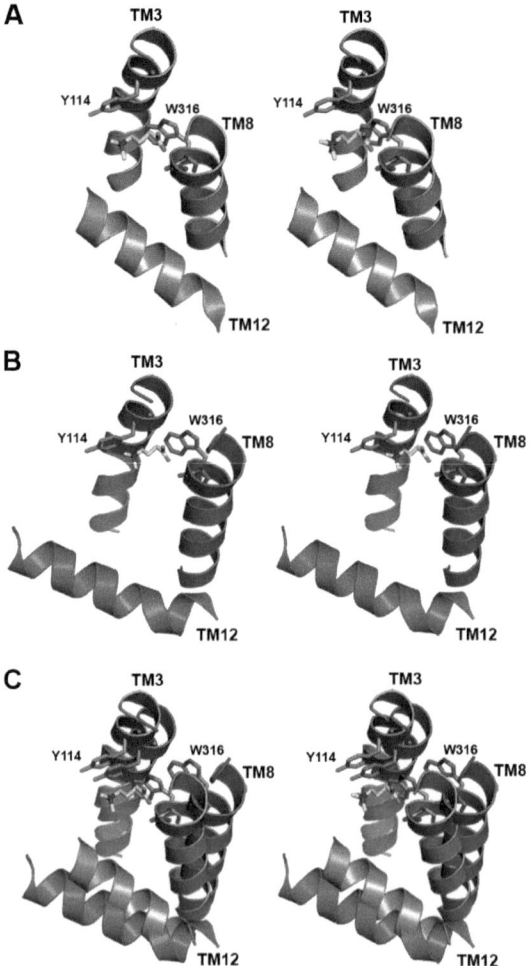

Figure 4-18 | Conformational changes in the transition from the inward-facing, open to the outward-facing, open conformation in CaiT

Stereo drawings comparing EcCaiT helices TM3, TM8 and TM12 with key residues in the inward-facing, open conformation (**A**, blue) to a model of EcCaiT in the outward-facing, open conformation (**B**, pink), based on the LeuT-Trp structure. In the model (**B**), the second, regulatory substrate-binding site on the extracellular side defined by residues Tyr114, Trp316 and the carbonyl oxygen of Gly315 remains intact, so that substrate is likely to remain bound throughout the transport cycle. (**C**) Superposition of **A** and **B**, indicating the helix movements that accompany the transition from the inside-open to the outside-open conformation on the extracellular side of CaiT.

4.5 Concluding remarks and outlook

At the beginning of this project little was known about the structure and mechanism of CaiT. K. R. Vinothkumar investigated in his PhD thesis the oligomeric state of EcCaiT and established the expression and a purification protocol that was refined for a successful 2D and initial 3D crystallization (Vinothkumar, 2005). The high *caiT* expression rate and the high protein yield was a prerequisite to start working on the 3D crystallization of CaiT. A crucial step during this work was to investigate CaiT homologs from different species in order to find one homolog that crystallized well and diffracted to high resolution. PmCaiT was the successful homolog with which the structure determination of CaiT was possible. The EcCaiT and StCaiT structures were subsequently solved with the well-refined PmCaiT model.

The 3D X-ray structure of the trimeric CaiT revealed interesting protein features. In the CaiT protomer two substrate-binding sites have been found. Binding studies with reconstituted protein showed that substrate binding in the antiporter CaiT is regulated in a heterotropic positive cooperated mechanism. Substrate/product antiport in CaiT is thus allosterically regulated. The two structures, PmCaiT and EcCaiT, also provide insights on how Na^+-independent substrate transport is achieved. In the EcCaiT structure an interesting and important interaction was found between the negatively charged carboxyl group of the substrate and the sulfur atom of a methionine in the central binding site. Although such an interaction has been postulated to occur in proteins (Pal and Chakrabarti, 2001), this does not seem to have been described before. This interaction allows the protein to coordinate the substrate without the need for additional binding partners. The same strategy is used at the equivalient position where H^+-dependent or Na^+-dependent transporter bind H^+ or Na^+ ions. In CaiT this position is taken by a positively charged arginine.

After analyzing the CaiT structure, point mutations were inserted into the protein and functional studies were performed to explain the mechanism of CaiT and to prove hypotheses that were made based on the structural information. All results

Discussion

taken together allowed us to propose a substrate/product exchange mechanism for CaiT.

The investigations of CaiT now need to focus on more detailed mechanistic problems. Many questions are open and need to be answered. One of those questions is the function of the CaiT trimer and if a protomer on its own would be functionally active. The complex substrate binding process needs to be resolved, i.e., the binding affinity of the regulatory binding site and the central binding site needs to be determined separately.

Regarding the L-carnitine production, it would be interesting to improve the substrate affinity and the tranport efficiency of CaiT.

It would also be very interesting to solve the CaiT structure in the outward-facing conformation and in the occluded state. Additional bioinformatic investigations with the structures of the three CaiT states would give further details of conformational changes that occur upon substrate binding and during the substrate/product exchange mechanism.

5 Literature

Abramson, J., and Wright, E.M. (2009). Structure and function of $Na^{(+)}$-symporters with inverted repeats. Curr Opin Struct Biol *19*, 425-432.

Adams, P.D., Afonine, P.V., Bunkoczi, G., Chen, V.B., Davis, I.W., Echols, N., Headd, J.J., Hung, L.W., Kapral, G.J., Grosse-Kunstleve, R.W., *et al.* (2010). PHENIX: a comprehensive Python-based system for macromolecular structure solution. Acta Crystallogr D Biol Crystallogr *66*, 213-221.

Adams, P.D., Gopal, K., Grosse-Kunstleve, R.W., Hung, L.W., Ioerger, T.R., McCoy, A.J., Moriarty, N.W., Pai, R.K., Read, R.J., Romo, T.D., *et al.* (2004). Recent developments in the PHENIX software for automated crystallographic structure determination. J Synchrotron Radiat *11*, 53-55.

Adams, P.D., Grosse-Kunstleve, R.W., Hung, L.W., Ioerger, T.R., McCoy, A.J., Moriarty, N.W., Read, R.J., Sacchettini, J.C., Sauter, N.K., and Terwilliger, T.C. (2002). PHENIX: building new software for automated crystallographic structure determination. Acta Crystallogr D Biol Crystallogr *58*, 1948-1954.

Angelidis, A.S., Smith, L.T., Hoffman, L.M., and Smith, G.M. (2002). Identification of opuC as a chill-activated and osmotically activated carnitine transporter in Listeria monocytogenes. Appl Environ Microbiol *68*, 2644-2650.

Baliarda, A., Robert, H., Jebbar, M., Blanco, C., and Le Marrec, C. (2003). Isolation and characterization of ButA, a secondary glycine betaine transport system operating in Tetragenococcus halophila. Curr Microbiol *47*, 347-351.

Barros, T., Royant, A., Standfuss, J., Dreuw, A., and Kühlbrandt, W. (2009). Crystal structure of plant light-harvesting complex shows the active, energy-transmitting state. EMBO J *28*, 298-306.

Bernal, V., Arense, P., Blatz, V., Mandrand-Berthelot, M.A., Canovas, M., and Iborra, J.L. (2008). Role of betaine:CoA ligase (CaiC) in the activation of betaines and the transfer of coenzyme A in Escherichia coli. J Appl Microbiol *105*, 42-50.

Blow, D.M. (2005). Outline of Crystallography for Biologists. Oxford University Press.

Blow, D.M., and Rossmann, M.G. (1961). The Single Isomorphous Replacement Method. Acta Cryst *14*, 1195-1202.

Boscari, A., Mandon, K., Dupont, L., Poggi, M.C., and Le Rudulier, D. (2002). BetS is a major glycine betaine/proline betaine transporter required for early osmotic adjustment in Sinorhizobium meliloti. J Bacteriol *184*, 2654-2663.

Bremer, J. (1983). Carnitine--metabolism and functions. Physiol Rev *63*, 1420-1480.

Bruschi, M., and Guerlesquin, F. (1988). Structure, function and evolution of bacterial ferredoxins. FEMS Microbiol Rev *4*, 155-175.

Buchet, A., Eichler, K., and Mandrand-Berthelot, M.A. (1998). Regulation of the carnitine pathway in Escherichia coli: investigation of the cai-fix divergent promoter region. J Bacteriol *180*, 2599-2608.

Burling, F.T., and Goldstein, B.M. (1993). A database study of nonbonded intramolecular sulfur-nucleophile contacts. Acta Crystallogr B *49 (Pt 4)*, 738-744.

Canaves, J.M., Page, R., Wilson, I.A., and Stevens, R.C. (2004). Protein biophysical properties that correlate with crystallization success in Thermotoga maritima: maximum clustering strategy for structural genomics. J Mol Biol *344*, 977-991.

Canovas, M., Torroglosa, T., Kleber, H.P., and Iborra, J.L. (2003). Effect of salt stress on crotonobetaine and D(+)-carnitine biotransformation into L(-)-carnitine by resting cells of Escherichia coli. J Basic Microbiol *43*, 259-268.

CCP4 (1994). The CCP4 suite: programs for protein crystallography. Acta Crystallogr D Biol Crystallogr *50*, 760-763.

Chen, C., and Beattie, G.A. (2008). Pseudomonas syringae BetT is a low-affinity choline transporter that is responsible for superior osmoprotection by choline over glycine betaine. J Bacteriol *190*, 2717-2725.

Chen, L., Oughtred, R., Berman, H.M., and Westbrook, J. (2004). TargetDB: a target registration database for structural genomics projects. Bioinformatics *20*, 2860-2862.

Chung, C.T., Niemela, S.L., and Miller, R.H. (1989). One-step preparation of competent Escherichia coli: transformation and storage of bacterial cells in the same solution. Proc Natl Acad Sci U S A *86*, 2172-2175.

Dauter, Z. (1999). Data-collection strategies. Acta Crystallogr D Biol Crystallogr *55*, 1703-1717.

Dauter, Z., Dauter, M., and Dodson, E. (2002). Jolly SAD. Acta Crystallogr D Biol Crystallogr *58*, 494-506.

Literature

de La Fortelle, E., and Bricogne, G. (1997). SHARP: a maximum-likelihood heavy-atom parameter refinement program for the MIR and MAD methods. Methods Enzymol *276*.

DeLano, W.L. (2002). The PyMOL Molecular Graphics System. DeLano Scientific LLC, San Carlos, CA, USA. https://www.pymol.org.

Dougherty, D.A. (1996). Cation-pi interactions in chemistry and biology: a new view of benzene, Phe, Tyr, and Trp. Science *271*, 163-168.

Dougherty, D.A. (2007). Cation-pi interactions involving aromatic amino acids. J Nutr *137*, 1504S-1508S; discussion 1516S-1517S.

Drenth, J. (2007). Principles of Protein X-Ray Crystallography. Springer Science + Business Media LLC *Third Edition*.

Durán, J.M., Peral, M.J., Calonge, M.L., and Ilundain, A.A. (2005). OCTN3: A Na^+-independent L-carnitine transporter in enterocytes basolateral membrane. J Cell Physiol *202*, 929-935.

Eichler, K., Bourgis, F., Buchet, A., Kleber, H.P., and Mandrand-Berthelot, M.A. (1994). Molecular characterization of the cai operon necessary for carnitine metabolism in Escherichia coli. Mol Microbiol *13*, 775-786.

Eichler, K., Buchet, A., Bourgis, F., Kleber, H.P., and Mandrand-Berthelot, M.A. (1995). The fix Escherichia coli region contains four genes related to carnitine metabolism. J Basic Microbiol *35*, 217-227.

Elimrani, I., Lahjouji, K., Seidman, E., Roy, M.J., Mitchell, G.A., and Qureshi, I. (2003). Expression and localization of organic cation/carnitine transporter OCTN2 in Caco-2 cells. Am J Physiol Gastrointest Liver Physiol *284*, G863-871.

Emsley, P., and Cowtan, K. (2004). Coot: model-building tools for molecular graphics. Acta Crystallogr D Biol Crystallogr *60*, 2126-2132.

Engemann, C., Elssner, T., Pfeifer, S., Krumbholz, C., Maier, T., and Kleber, H.P. (2005). Identification and functional characterisation of genes and corresponding enzymes involved in carnitine metabolism of Proteus sp. Arch Microbiol *183*, 176-189.

Esnouf, R.M. (1999). Further additions to MolScript version 1.4, including reading and contouring of electron-density maps. Acta Crystallogr D Biol Crystallogr *55*, 938-940.

Evans, P. (2006). Scaling and assessment of data quality. Acta Crystallogr D Biol Crystallogr 62, 72-82.

Faham, S., Watanabe, A., Besserer, G.M., Cascio, D., Specht, A., Hirayama, B.A., Wright, E.M., and Abramson, J. (2008). The crystal structure of a sodium galactose transporter reveals mechanistic insights into Na^+/sugar symport. Science 321, 810-814.

Fan, X., Pericone, C.D., Lysenko, E., Goldfine, H., and Weiser, J.N. (2003). Multiple mechanisms for choline transport and utilization in Haemophilus influenzae. Mol Microbiol 50, 537-548.

Fang, Y., Jayaram, H., Shane, T., Kolmakova-Partensky, L., Wu, F., Williams, C., Xiong, Y., and Miller, C. (2009). Structure of a prokaryotic virtual proton pump at 3.2 A resolution. Nature 460, 1040-1043.

Farwick, M., Siewe, R.M., and Kramer, R. (1995). Glycine betaine uptake after hyperosmotic shift in Corynebacterium glutamicum. J Bacteriol 177, 4690-4695.

Fenn, T.D., Ringe, D., and Petsko, G.A. (2002). POVScript+: a program for model and data visualization using persistence of vision ray-tracing. J Appl Crystallogr 36, 944-947.

Forrest, L.R., and Rudnick, G. (2009). The rocking bundle: a mechanism for ion-coupled solute flux by symmetrical transporters. Physiology (Bethesda) 24, 377-386.

Forrest, L.R., Zhang, Y.W., Jacobs, M.T., Gesmonde, J., Xie, L., Honig, B.H., and Rudnick, G. (2008). Mechanism for alternating access in neurotransmitter transporters. Proc Natl Acad Sci U S A 105, 10338-10343.

Fox, J.D., and Robyt, J.F. (1991). Miniaturization of three carbohydrate analyses using a microsample plate reader. Anal Biochem 195, 93-96.

Gallivan, J.P., and Dougherty, D.A. (1999). Cation-pi interactions in structural biology. Proc Natl Acad Sci U S A 96, 9459-9464.

Gao, X., Lu, F., Zhou, L., Dang, S., Sun, L., Li, X., Wang, J., and Shi, Y. (2009). Structure and mechanism of an amino acid antiporter. Science 324, 1565-1568.

Gao, X., Zhou, L., Jiao, X., Lu, F., Yan, C., Zeng, X., Wang, J., and Shi, Y. (2010). Mechanism of substrate recognition and transport by an amino acid antiporter. Nature 463, 828-832.

Garman, E.F., and Doublié, S. (2003). Cryocooling of macromolecular crystals: optimization methods. Methods Enzymol 368, 188-216.

Garman, E.F., and Schneider, T.R. (1997). Macromolecular cryocrystallography. J Appl Crystallography *30*, 211-237.

Glasel, J.A. (1995). Validity of nucleic acid purities monitored by 260nm/280nm absorbance ratios. Biotechniques *18*, 62-63.

González, A. (2003). Optimizing data collection for structure determination. Acta Crystallogr D Biol Crystallogr *59*, 1935-1942.

González, A., Pedelacq, J., Sola, M., Gomis-Ruth, F.X., Coll, M., Samama, J., and Benini, S. (1999). Two-wavelength MAD phasing: in search of the optimal choice of wavelengths. Acta Crystallogr D Biol Crystallogr *55*, 1449-1458.

Haardt, M., and Bremer, E. (1996). Use of phoA and lacZ fusions to study the membrane topology of ProW, a component of the osmoregulated ProU transport system of Escherichia coli. J Bacteriol *178*, 5370-5381.

Haas, D.J. (1968). Preliminary studies on the denaturation of cross-linked lysozyme crystals. Biophys J *8*, 549-555.

Hendrickson, W.A., and Lattman, E.E. (1970). Representation of Phase Probability Distributions for Simplified Combination of Independent Phase Information. Acta Crystallogr B *26*, 136-143.

Horne, D.W., and Broquist, H.P. (1973). Role of lysine and -N-trimethyllysine in carnitine biosynthesis. I. Studies in Neurospora crassa. J Biol Chem *248*, 2170-2175.

Jablonski, A. (1935). Über den Mechanismus der Photolumineszenz von Farbstoffphophoren. Zeitschrift für Physik *94*, 38-46.

Jardetzky, O. (1966). Simple allosteric model for membrane pumps. Nature *211*, 969-970.

Jones, T.A., Zou, J.Y., Cowan, S.W., and Kjeldgaard, M. (1991). Improved methods for building protein models in electron density maps and the location of errors in these models. Acta Crystallogr A *47 (Pt 2)*, 110-119.

Jung, H., Buchholz, M., Clausen, J., Nietschke, M., Revermann, A., Schmid, R., and Jung, K. (2002). CaiT of Escherichia coli, a new transporter catalyzing L-carnitine/gamma -butyrobetaine exchange. J Biol Chem *277*, 39251-39258.

Jung, H., Jung, K., and Kleber, H.P. (1989). Purification and properties of carnitine dehydratase from Escherichia coli--a new enzyme of carnitine metabolization. Biochim Biophys Acta *1003*, 270-276.

Jung, H., Jung, K., and Kleber, H.P. (1990a). L-carnitine metabolization and osmotic stress response in Escherichia coli. J Basic Microbiol *30*, 409-413.

Jung, H., Jung, K., and Kleber, H.P. (1990b). L-carnitine uptake by Escherichia coli. J Basic Microbiol *30*, 507-514.

Jung, K., Jung, H., and Kleber, H.P. (1987). Regulation of L-carnitine metabolism in Escherichia coli. J Basic Microbiol *27*, 131-137.

Kabsch, W. (1993). Automatic Processing of Rotation Diffraction Data from Crystals of Initially Unknown Symmetry and Cell Constants. J Appl Crystallogr *26*, 795-800.

Kappes, R.M., Kempf, B., and Bremer, E. (1996). Three transport systems for the osmoprotectant glycine betaine operate in Bacillus subtilis: characterization of OpuD. J Bacteriol *178*, 5071-5079.

Kleber, H.P. (1997). Bacterial carnitine metabolism. FEMS Microbiol Lett *147*, 1-9.

Ko, R., Smith, L.T., and Smith, G.M. (1994). Glycine betaine confers enhanced osmotolerance and cryotolerance on Listeria monocytogenes. J Bacteriol *176*, 426-431.

Koshland, D.E. (1958). Application of a Theory of Enzyme Specificity to Protein Synthesis. Proc Natl Acad Sci U S A *44*, 98-104.

Koshland, D.E., Jr. (1973). [Induced conformational changes in enzyme control (author's transl)]. Seikagaku *45*, 941-950.

Koshland, D.E., Jr., Nemethy, G., and Filmer, D. (1966). Comparison of experimental binding data and theoretical models in proteins containing subunits. Biochemistry *5*, 365-385.

Krämer, R., and Morbach, S. (2004). BetP of Corynebacterium glutamicum, a transporter with three different functions: betaine transport, osmosensing, and osmoregulation. Biochim Biophys Acta *1658*, 31-36.

Krämer, R., and Ziegler, C. (2009). Regulative interactions of the osmosensing C-terminal domain in the trimeric glycine betaine transporter BetP from Corynebacterium glutamicum. Biol Chem *390*, 685-691.

Krishnamurthy, H., Piscitelli, C.L., and Gouaux, E. (2009). Unlocking the molecular secrets of sodium-coupled transporters. Nature *459*, 347-355.

Krissinel, E., and Henrick, K. (2004). Secondary-structure matching (SSM), a new tool for fast protein structure alignment in three dimensions. Acta Crystallogr D Biol Crystallogr *60*, 2256-2268.

Laemmli, U.K. (1970). Cleavage of structural proteins during the assembly of the head of bacteriophage T4. Nature *227*, 680-685.

Lamark, T., Kaasen, I., Eshoo, M.W., Falkenberg, P., McDougall, J., and Strom, A.R. (1991). DNA sequence and analysis of the bet genes encoding the osmoregulatory choline-glycine betaine pathway of Escherichia coli. Mol Microbiol *5*, 1049-1064.

Laskowski, R.A., Moss, D.S., and Thornton, J.M. (1993). Main-chain bond lengths and bond angles in protein structures. J Mol Biol *231*, 1049-1067.

Le Rudulier, D., Strom, A.R., Dandekar, A.M., Smith, L.T., and Valentine, R.C. (1984). Molecular biology of osmoregulation. Science *224*, 1064-1068.

Leslie, A.G.W. (1992). Recent changes to the MOSFLM package for processing film and image plate data. Joint CCP4 + ESF-EAMCB Newsletter on Protein Crystallography *26*.

Lewis, B.A., and Engelman, D.M. (1983). Lipid bilayer thickness varies linearly with acyl chain length in fluid phosphatidylcholine vesicles. J Mol Biol *166*, 211-217.

Ly, A., Henderson, J., Lu, A., Culham, D.E., and Wood, J.M. (2004). Osmoregulatory systems of Escherichia coli: identification of betaine-carnitine-choline transporter family member BetU and distributions of betU and trkG among pathogenic and nonpathogenic isolates. J Bacteriol *186*, 296-306.

MacMillan, S.V., Alexander, D.A., Culham, D.E., Kunte, H.J., Marshall, E.V., Rochon, D., and Wood, J.M. (1999). The ion coupling and organic substrate specificities of osmoregulatory transporter ProP in Escherichia coli. Biochim Biophys Acta *1420*, 30-44.

Mahling, S., Asmus, K.D., Glass, R.S., Hojjatie, M., and Wilson, G.S. (1987). Neighboring group participation in radicals: pulse radiolysis studies on radicals with sulfur-oxygen interaction. J Org Chem *52*, 3717-3724.

McCoy, A.J., Grosse-Kunstleve, R.W., Adams, P.D., Winn, M.D., Storoni, L.C., and Read, R.J. (2007). Phaser crystallographic software. J Appl Crystallogr *40*, 658-674.

McPherson, A. (1999). Crystallization of biological macromolecules. Cold Spring Harbor Laboratory Press.

McPherson, A. (2004). Introduction to protein crystallization. Methods *34*, 254-265.

Messerschmidt, A. (2007). X-Ray Crystallography of Biomacromolecules. A Practical Guide. Wiley-VCH Verlag GmbH & Co KGaA.

Moffatt, B.A., and Studier, F.W. (1987). T7 lysozyme inhibits transcription by T7 RNA polymerase. Cell *49*, 221-227.

Morbach, S., and Krämer, R. (2005). Structure and function of the betaine uptake system BetP of Corynebacterium glutamicum: strategies to sense osmotic and chill stress. J Mol Microbiol Biotechnol *10*, 143-153.

Mullis, K.B., and Faloona, F.A. (1987). Specific synthesis of DNA in vitro via a polymerase-catalyzed chain reaction. Methods Enzymol *155*, 335-350.

Nave, C., and Garman, E.F. (2005). Towards an understanding of radiation damage in cryocooled macromolecular crystals. J Synchrotron Radiat *12*, 257-260.

Noble, C.L., Nimmo, E.R., Drummond, H., Ho, G.T., Tenesa, A., Smith, L., Anderson, N., Arnott, I.D., and Satsangi, J. (2005). The contribution of OCTN1/2 variants within the IBD5 locus to disease susceptibility and severity in Crohn's disease. Gastroenterology *129*, 1854-1864.

Norvell, J.C., and Machalek, A.Z. (2000). Structural genomics programs at the US National Institute of General Medical Sciences. Nat Struct Biol *7 Suppl*, 931.

Obon, J.M., Maiquez, J.R., Canovas, M., Kleber, H.P., and Iborra, J.L. (1999). High-density Escherichia coli cultures for continuous L(-)-carnitine production. Appl Microbiol Biotechnol *51*, 760-764.

Pal, D., and Chakrabarti, P. (2001). Non-hydrogen bond interactions involving the methionine sulfur atom. J Biomol Struct Dyn *19*, 115-128.

Palsdottir, H., and Hunte, C. (2004). Lipids in membrane protein structures. Biochim Biophys Acta *1666*, 2-18.

Peltekova, V.D., Wintle, R.F., Rubin, L.A., Amos, C.I., Huang, Q., Gu, X., Newman, B., Van Oene, M., Cescon, D., Greenberg, G., *et al.* (2004). Functional variants of OCTN cation transporter genes are associated with Crohn disease. Nat Genet *36*, 471-475.

Peter, H., Burkovski, A., and Kramer, R. (1996). Isolation, characterization, and expression of the Corynebacterium glutamicum betP gene, encoding the transport system for the compatible solute glycine betaine. J Bacteriol *178*, 5229-5234.

Quick, M., Winther, A.M., Shi, L., Nissen, P., Weinstein, H., and Javitch, J.A. (2009). Binding of an octylglucoside detergent molecule in the second substrate (S2) site of LeuT establishes an inhibitor-bound conformation. Proc Natl Acad Sci U S A *106*, 5563-5568.

Ramachandran, G.N., Ramakrishnan, C., and Sasisekharan, V. (1963). Stereochemistry of polypeptide chain configurations. J Mol Biol *7*, 95-99.

Ramsay, R.R., Gandour, R.D., and van der Leij, F.R. (2001). Molecular enzymology of carnitine transfer and transport. Biochim Biophys Acta *1546*, 21-43.

Rebouche, C.J., and Seim, H. (1998). Carnitine metabolism and its regulation in microorganisms and mammals. Annu Rev Nutr *18*, 39-61.

Ressl, S., Terwisscha van Scheltinga, A.C., Vonrhein, C., Ott, V., and Ziegler, C. (2009). Molecular basis of transport and regulation in the $Na^{(+)}$/betaine symporter BetP. Nature *458*, 47-52.

Reyes, N., Ginter, C., and Boudker, O. (2009). Transport mechanism of a bacterial homologue of glutamate transporters. Nature *462*, 880-885.

Ricard, J., and Cornish-Bowden, A. (1987). Co-operative and allosteric enzymes: 20 years on. Eur J Biochem *166*, 255-272.

Rosenfield, J., R.E., Parthasarathy, R., and Dunitz, J.D. (1977). Directional Preferences of Nonbonded Atomic Contacts with Divalent Sulfur. 1. Electrophiles and Nucleophiles. J Am Chem Soc *99*, 4860-4862.

Rosenstein, R., Futter-Bryniok, D., and Gotz, F. (1999). The choline-converting pathway in Staphylococcus xylosus C2A: genetic and physiological characterization. J Bacteriol *181*, 2273-2278.

Rossmann, M.G., and Blow, D.M. (1962). The detection of sub-units within the crystallographic asymmetric unit. Acta Crystallogr *15*, 24-31.

Saiki, R.K., Gelfand, D.H., Stoffel, S., Scharf, S.J., Higuchi, R., Horn, G.T., Mullis, K.B., and Erlich, H.A. (1988). Primer-directed enzymatic amplification of DNA with a thermostable DNA polymerase. Science *239*, 487-491.

Schaffner, W., and Weissmann, C. (1973). A rapid, sensitive, and specific method for the determination of protein in dilute solution. Anal Biochem *56*, 502-514.

Schagger, H., and von Jagow, G. (1991). Blue native electrophoresis for isolation of membrane protein complexes in enzymatically active form. Anal Biochem *199*, 223-231.

Schneider, T.R., and Sheldrick, G.M. (2002). Substructure solution with SHELXD. Acta Crystallogr D Biol Crystallogr 58, 1772-1779.

Seim, H., and Kleber, H.P. (1988). Synthesis of L(-)-carnitine by hydration of crotonobetaine by enterobacteria. Appl Microbiol Biotechnol 27, 538-544.

Shaffer, P.L., Goehring, A., Shankaranarayanan, A., and Gouaux, E. (2009). Structure and mechanism of a Na^+-independent amino acid transporter. Science 325, 1010-1014.

Shi, L., Quick, M., Zhao, Y., Weinstein, H., and Javitch, J.A. (2008). The mechanism of a neurotransmitter:sodium symporter--inward release of Na^+ and substrate is triggered by substrate in a second binding site. Mol Cell 30, 667-677.

Shimamura, T., Weyand, S., Beckstein, O., Rutherford, N.G., Hadden, J.M., Sharples, D., Sansom, M.S., Iwata, S., Henderson, P.J., and Cameron, A.D. (2010). Molecular basis of alternating access membrane transport by the sodium-hydantoin transporter Mhp1. Science 328, 470-473.

Singh, S.K., Piscitelli, C.L., Yamashita, A., and Gouaux, E. (2008). A competitive inhibitor traps LeuT in an open-to-out conformation. Science 322, 1655-1661.

Slabinski, L., Jaroszewski, L., Rodrigues, A.P., Rychlewski, L., Wilson, I.A., Lesley, S.A., and Godzik, A. (2007a). The challenge of protein structure determination--lessons from structural genomics. Protein Sci 16, 2472-2482.

Slabinski, L., Jaroszewski, L., Rychlewski, L., Wilson, I.A., Lesley, S.A., and Godzik, A. (2007b). XtalPred: a web server for prediction of protein crystallizability. Bioinformatics 23, 3403-3405.

Sleator, R.D., Gahan, C.G., Abee, T., and Hill, C. (1999). Identification and disruption of BetL, a secondary glycine betaine transport system linked to the salt tolerance of Listeria monocytogenes LO28. Appl Environ Microbiol 65, 2078-2083.

Smith, L.T. (1996). Role of osmolytes in adaptation of osmotically stressed and chill-stressed Listeria monocytogenes grown in liquid media and on processed meat surfaces. Appl Environ Microbiol 62, 3088-3093.

Steger, R., Weinand, M., Kramer, R., and Morbach, S. (2004). LcoP, an osmoregulated betaine/ectoine uptake system from Corynebacterium glutamicum. FEBS Lett 573, 155-160.

Studier, F.W. (1991). Use of bacteriophage T7 lysozyme to improve an inducible T7 expression system. J Mol Biol 219, 37-44.

Studier, F.W. (2005). Protein production by auto-induction in high density shaking cultures. Protein Expr Purif *41*, 207-234.

Studts, J.M., and Fox, B.G. (1999). Application of fed-batch fermentation to the preparation of isotopically labeled or selenomethionyl-labeled proteins. Protein Expr Purif *16*, 109-119.

Styrvold, O.B., Falkenberg, P., Landfald, B., Eshoo, M.W., Bjornsen, T., and Strom, A.R. (1986). Selection, mapping, and characterization of osmoregulatory mutants of Escherichia coli blocked in the choline-glycine betaine pathway. J Bacteriol *165*, 856-863.

Tamai, I., China, K., Sai, Y., Kobayashi, D., Nezu, J., Kawahara, E., and Tsuji, A. (2001). $Na^{(+)}$-coupled transport of L-carnitine via high-affinity carnitine transporter OCTN2 and its subcellular localization in kidney. Biochim Biophys Acta *1512*, 273-284.

Tamai, I., Ohashi, R., Nezu, J.I., Sai, Y., Kobayashi, D., Oku, A., Shimane, M., and Tsuji, A. (2000). Molecular and functional characterization of organic cation/carnitine transporter family in mice. J Biol Chem *275*, 40064-40072.

Tamai, I., Yabuuchi, H., Nezu, J., Sai, Y., Oku, A., Shimane, M., and Tsuji, A. (1997). Cloning and characterization of a novel human pH-dependent organic cation transporter, OCTN1. FEBS Lett *419*, 107-111.

Tang, L., Bai, L., Wang, W.H., and Jiang, T. (2010). Crystal structure of the carnitine transporter and insights into the antiport mechanism. Nat Struct Mol Biol *17*, 492-496.

Tanphaichitr, V., and Broquist, H.P. (1973). Role of lysine and -N-trimethyllysine in carnitine biosynthesis. II. Studies in the rat. J Biol Chem *248*, 2176-2181.

Tein, I. (2003). Carnitine transport: pathophysiology and metabolism of known molecular defects. J Inherit Metab Dis *26*, 147-169.

Terwilliger, T.C. (2000). Maximum-likelihood density modification. Acta Crystallogr D Biol Crystallogr *56*, 965-972.

Terwilliger, T.C. (2001). Map-likelihood phasing. Acta Crystallogr D Biol Crystallogr *57*, 1763-1775.

Terwilliger, T.C. (2004). Using prime-and-switch phasing to reduce model bias in molecular replacement. Acta Crystallogr D Biol Crystallogr *60*, 2144-2149.

Terwilliger, T.C., and Berendzen, J. (1999). Automated MAD and MIR structure solution. Acta Crystallogr D Biol Crystallogr 55, 849-861.

Tondervik, A., and Strom, A.R. (2007). Membrane topology and mutational analysis of the osmotically activated BetT choline transporter of Escherichia coli. Microbiology 153, 803-813.

Tsai, M.H., and Saier, M.H., Jr. (1995). Phylogenetic characterization of the ubiquitous electron transfer flavoprotein families ETF-alpha and ETF-beta. Res Microbiol 146, 397-404.

Urbani, A., and Warne, T. (2005). A colorimetric determination for glycosidic and bile salt-based detergents: applications in membrane protein research. Anal Biochem 336, 117-124.

Vaz, F.M., and Wanders, R.J. (2002). Carnitine biosynthesis in mammals. Biochem J 361, 417-429.

Verheul, A., Wouters, J.A., Rombouts, F.M., and Abee, T. (1998). A possible role of ProP, ProU and CaiT in osmoprotection of Escherichia coli by carnitine. J Appl Microbiol 85, 1036-1046.

Vermeulen, V., and Kunte, H.J. (2004). Marinococcus halophilus DSM 20408T encodes two transporters for compatible solutes belonging to the betaine-carnitine-choline transporter family: identification and characterization of ectoine transporter EctM and glycine betaine transporter BetM. Extremophiles 8, 175-184.

Vinothkumar, K.R. (2005). Structural Studies of Membrane Transport Proteins. PhD Thesis, JW-Goethe University, Frankfurt am Main.

Vinothkumar, K.R., Raunser, S., Jung, H., and Kühlbrandt, W. (2006). Oligomeric structure of the carnitine transporter CaiT from Escherichia coli. J Biol Chem 281, 4795-4801.

von Heijne, G., and Gavel, Y. (1988). Topogenic signals in integral membrane proteins. Eur J Biochem 174, 671-678.

Vonrhein, C., Blanc, E., Roversi, P., and Bricogne, G. (2007). Automated structure solution with autoSHARP. Methods Mol Biol 364, 215-230.

Walt, A., and Kahn, M.L. (2002). The fixA and fixB genes are necessary for anaerobic carnitine reduction in Escherichia coli. J Bacteriol 184, 4044-4047.

Warburg, O., and Christian, W. (1942). Isolierung und Kristallisation des Gärungsferments Enolase. Biochem Z 310, 384-421.

Literature

Weik, M., and Colletier, J.P. (2010). Temperature-dependent macromolecular X-ray crystallography. Acta Crystallogr D Biol Crystallogr *66*, 437-446.

Weyand, S., Shimamura, T., Yajima, S., Suzuki, S., Mirza, O., Krusong, K., Carpenter, E.P., Rutherford, N.G., Hadden, J.M., O'Reilly, J., *et al.* (2008). Structure and molecular mechanism of a nucleobase-cation-symport-1 family transporter. Science *322*, 709-713.

Wu, X., George, R.L., Huang, W., Wang, H., Conway, S.J., Leibach, F.H., and Ganapathy, V. (2000). Structural and functional characteristics and tissue distribution pattern of rat OCTN1, an organic cation transporter, cloned from placenta. Biochim Biophys Acta *1466*, 315-327.

Wu, X., Huang, W., Prasad, P.D., Seth, P., Rajan, D.P., Leibach, F.H., Chen, J., Conway, S.J., and Ganapathy, V. (1999). Functional characteristics and tissue distribution pattern of organic cation transporter 2 (OCTN2), an organic cation/carnitine transporter. J Pharmacol Exp Ther *290*, 1482-1492.

Yamashita, A., Singh, S.K., Kawate, T., Jin, Y., and Gouaux, E. (2005). Crystal structure of a bacterial homologue of Na^+/Cl^--dependent neurotransmitter transporters. Nature *437*, 215-223.

Yancey, P.H., Clark, M.E., Hand, S.C., Bowlus, R.D., and Somero, G.N. (1982). Living with water stress: evolution of osmolyte systems. Science *217*, 1214-1222.

Yernool, D., Boudker, O., Jin, Y., and Gouaux, E. (2004). Structure of a glutamate transporter homologue from Pyrococcus horikoshii. Nature *431*, 811-818.

Zhou, Z., Zhen, J., Karpowich, N.K., Goetz, R.M., Law, C.J., Reith, M.E., and Wang, D.N. (2007). LeuT-desipramine structure reveals how antidepressants block neurotransmitter reuptake. Science *317*, 1390-1393.

Ziegler, C., Morbach, S., Schiller, D., Kramer, R., Tziatzios, C., Schubert, D., and Kühlbrandt, W. (2004). Projection structure and oligomeric state of the osmoregulated sodium/glycine betaine symporter BetP of Corynebacterium glutamicum. J Mol Biol *337*, 1137-1147.

Literature

6 Appendix

6.1 Protein constructs

PmCaiT

UniProtKB entry name: **CaiT_PROSL**   UniProtKB accession no.: **P59334**

```
>sp|P59334|CAIT_PROSL    L-carnitine/gamma-butyrobetaine    antiporter
OS=Proteus sp. (strain LE138) GN=caiT PE=3 SV=1
MSKDNKKAGIEPKVFFPPLIIVGILCWLTVRDLDASNEVINAVFSYVTNVWGWAFEWYMV
IMFGGWFWLVFGRYAKKRLGDEKPEFSTASWIFMMFASCTSAAVLFWGSIEIYYYISSPP
FGMEGYSAPAKEIGLAYSLFHWGPLPWATYSFLSVAFAYFFFVRKMEVIRPSSTLTPLVG
EKHVNGLFGTVVDNFYLVALILAMGTSLGLATPLVTECIQYLFGIPHTLQLDAIIISCWI
LLNAICVAFGLQKGVKIASDVRTYLSFLMLGWVFIVGGASFIVNYFTDSVGTLLMYMPRM
LFYTDPIGKGGFPQAWTVFYWAWWVIYAIQMSIFLARISKGRTVRELCLGMVSGLTAGTW
LIWTYSGGNTLQLIDQNILNIPQLIDQYGVPRAIIETWAALPLSTATMWGFFILCFIATV
TLINACSYTLAMSTCRSMKEGAEPPLLVRIGWSVLVGIIGIILLALGGLKPIQTAIIAGG
CPLFFVNIMVTLSFIKDAKVHWKD
```

EcCaiT

UniProtKB entry name: **CaiT_ECOLI**   UniProtKB accession no.: **P31553**

```
>sp|P31553|CAIT_ECOLI    L-carnitine/gamma-butyrobetaine    antiporter
OS=Escherichia coli (strain K12) GN=caiT PE=1 SV=2
MKNEKRKTGIEPKVFFPPLIIVGILCWLTVRDLDAANVVINAVFSYVTNVWGWAFEWYMV
VMLFGWFWLVFGPYAKKRLGNEPPEFSTASWIFMMFASCTSAAVLFWGSIEIYYYISTPP
FGLEPNSTGAKELGLAYSLFHWGPLPWATYSFLSVAFAYFFFVRKMEVIRPSSTLVPLVG
EKHAKGLFGTIVDNFYLVALIFAMGTSLGLATPLVTECMQWLFGIPHTLQLDAIIITCWI
ILNAICVACGLQKGVRIASDVRSYLSFLMLGWVFIVSGASFIMNYFTDSVGMLLMYLPRM
LFYTDPIAKGGFPQGWTVFYWAWWVIYAIQMSIFLARISRGRTVRELCFGMVLGLTASTW
ILWTVLGSNTLLLIDKNIINIPNLIEQYGVARAIIETWAALPLSTATMWGFFILCFIATV
TLVNACSYTLAMSTCREVRDGEEPPLLVRIGWSILVGIIGIVLLALGGLKPIQTAIIAGG
CPLFFVNIMVTLSFIKDAKQNWKD
```

Appendix

StCaiT

UniProtKB entry name: **CaiT_SALTI**   UniProtKB accession no.: **Q8Z9L1**

```
>sp|Q8Z9L1|CAIT_SALTI   L-carnitine/gamma-butyrobetaine   antiporter
OS=Salmonella typhi GN=caiT PE=3 SV=1
MKNEKRKSGIEPKVFFPLLIIVGILCWLTVRDLDAANVVINAVFSYVTNVWGWAFEWYMV
VMLFSWFWLVFGPYAKKRLGDEKPEFSTASWIFMMFASCTSAAVLFWGSIEIYYYISTPP
FGLEPNSTGAKEIGLAYSLFHWGPLPWATYSFLSVAFAYFFFVRKMDVIRPSSTLVPLVG
EKHAKGLFSTIVDNFYLVALIFAMGTSLGLATPLVTECMQWLFGIPHTLQLDAIIITCWI
ILNAICVACGLQKGVRIASDVRSYLSFLMLGWVFIVSGASFIMNYFTDSVGMLLMHLPRM
LFYTDAIGKGGFPQGWTVFYWAWWVIYAIQMSIFLARISRGRTVRELCFGMVMGLTASTW
ILWTVLGSNTLLLMDKNILNIPQLIEQHGVARAIIETWAALPLSTATMWGFFILCFIATV
TLINACSYTLAMSTCREVRDGEEPPLLVRIGWSVLVGIIGIVLLALGGLKPIQTAIIAGG
CPLFFVNIMVTLSFIKDAKVHWKDK
```

Appendix

6.2 Vector System

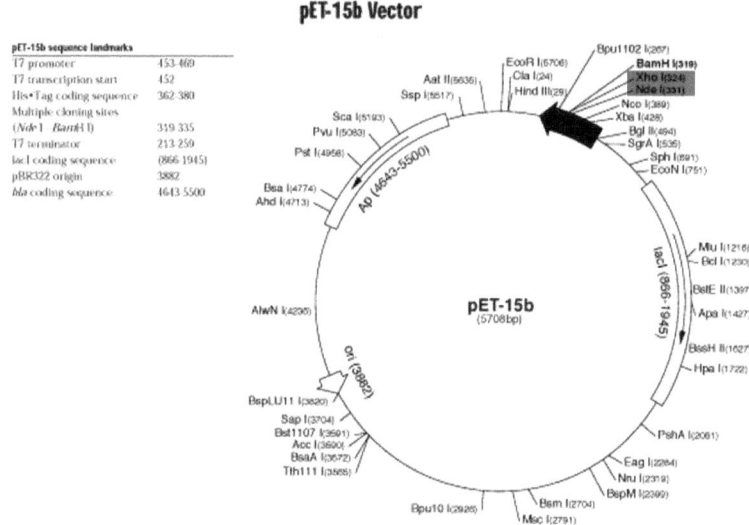

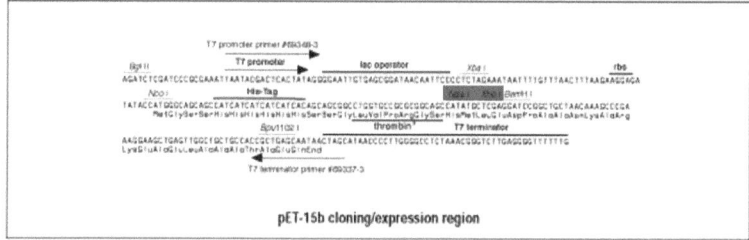

Figure 6-1 | pET-15b vector map

The Ec*caiT*, Pm*caiT* or St*caiT* gene was cloned into the pET-15b (Novagen) vector, using the restriction enzymes *Nde*I and *Xho*I (red box).

Appendix

6.3 Sequence Alignment

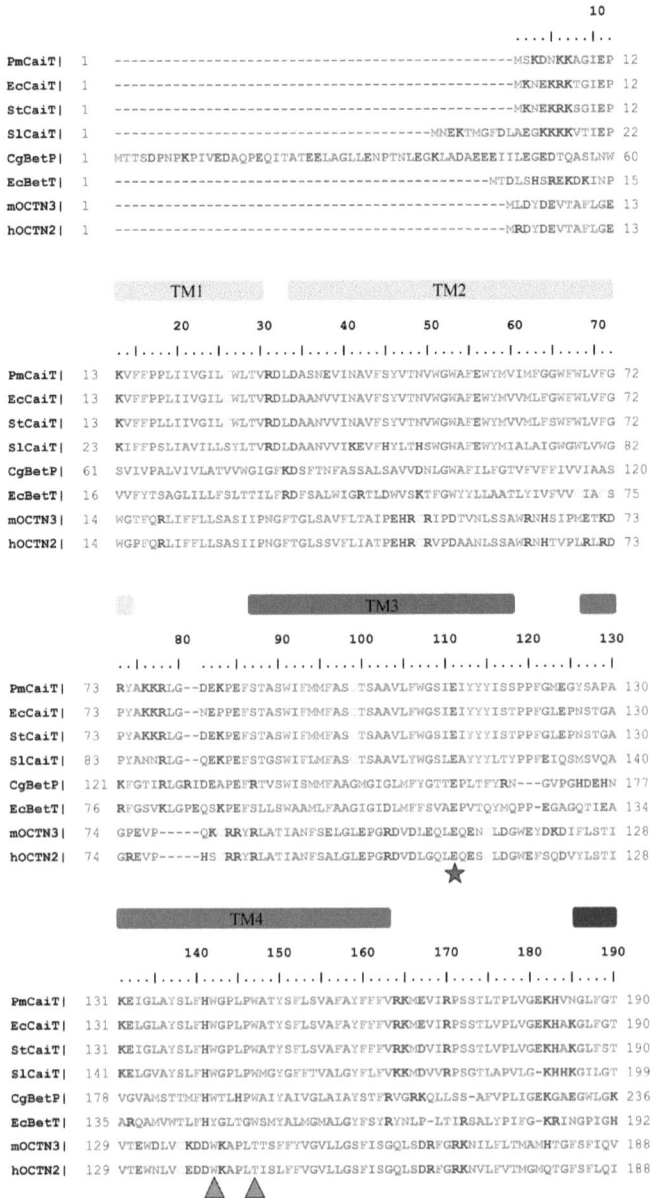

214

Appendix

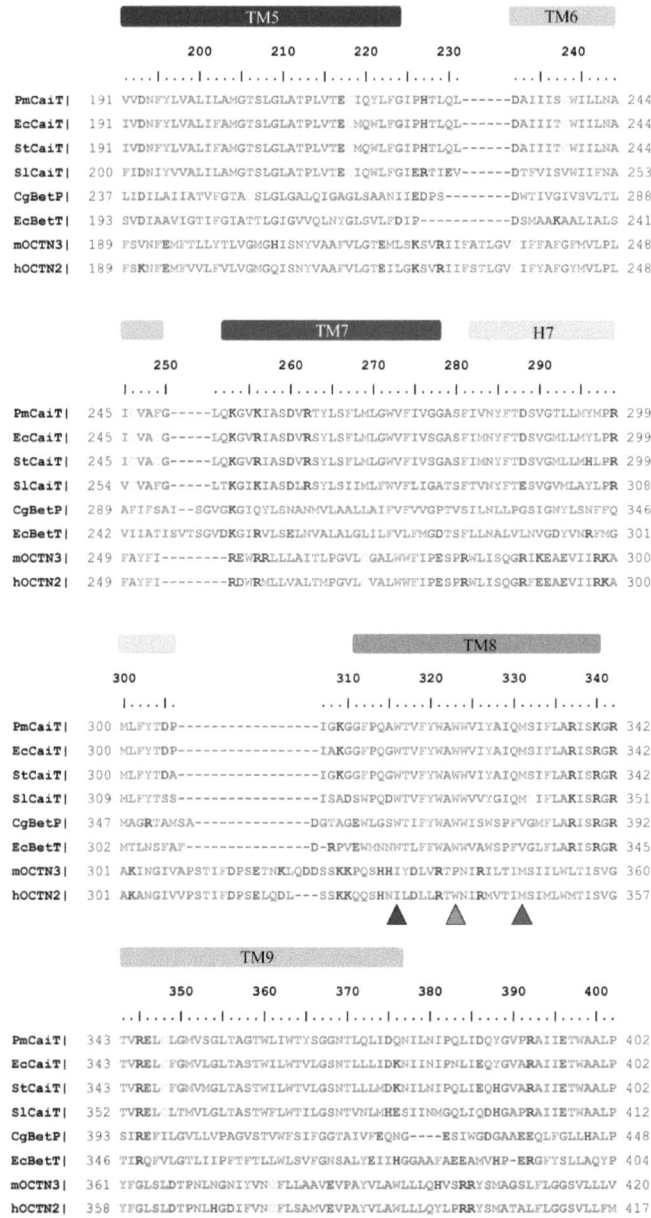

Appendix

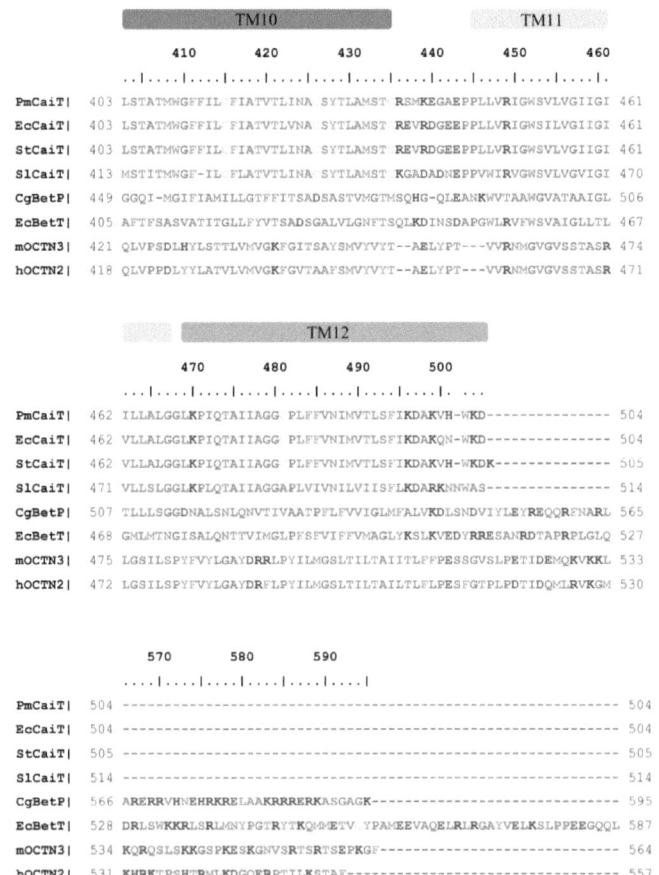

Figure 6-2 | Sequence Alingment

Amino acid sequence alignment of CaiT from *Proteus mirabilis* (PmCaiT), *E. coli* (EcCaiT), *Salmonella typhimorium* (StCaiT), *Shewanella loihica* (SlCaiT), BetP from *Corynebacterium glutamicum* (CgBetP), *E.coli* BetT (EcBetT), and organic cation transporters from mouse (mOCTN3) and human (hOCTN2). Red triangle: Met331 which coordinates the carboxyl group of the bound substrate in the transport site of CaiT. Orange triangles: fully or partly conserved hydrophobic residues in the central transport site. Blue triangle: Trp316 in the regulatory site of CaiT. Red asterisk: Glu111, which coordinates the network of hydrogen bonds linking the two inverted repeats in the inside-open conformation of CaiT. Residues 588 – 677 of EcBetT were omitted.

Appendix

6.4 Crystal packing

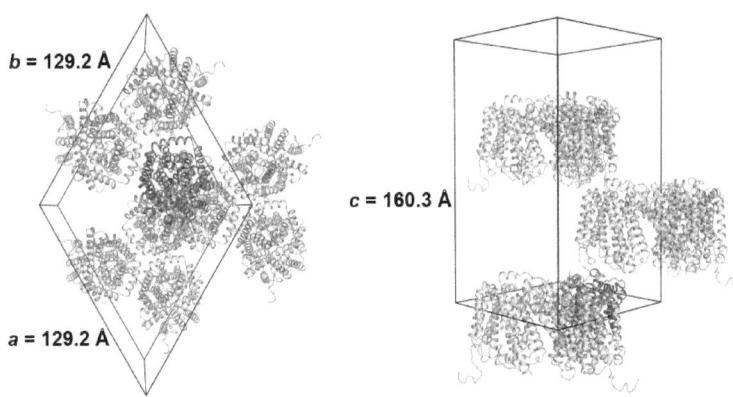

Figure 6-3 | Unit cell packing of PmCaiT

Packing of PmCaiT within the unit cell $H3$ with the cell parameters $a = b = 129.2$ Å, $c = 160.3$ Å, $\alpha = \beta = 90°$, $\gamma = 120°$.

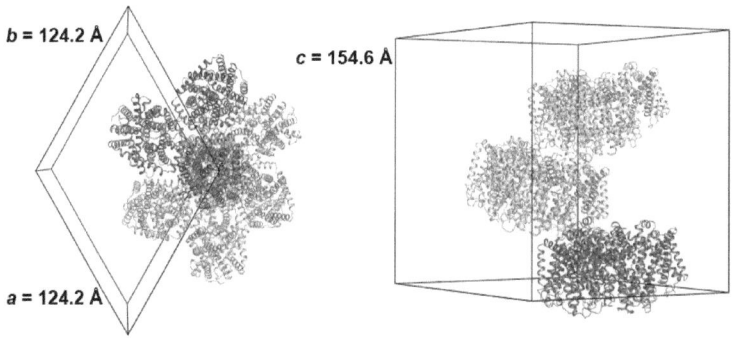

Figure 6-4 | Unit cell packing of EcCaiT

Packing of EcCaiT within the unit cell $P3_2$ with the cell parameters $a = b = 124.2$ Å, $c = 154.6$ Å, $\alpha = \beta = 90°$, $\gamma = 120°$.

217

Appendix

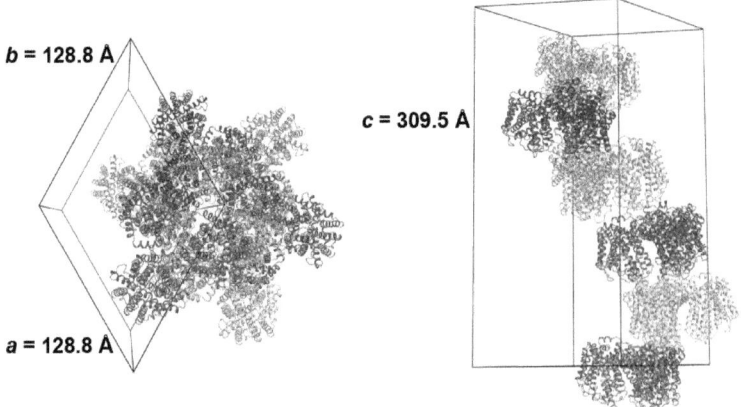

Figure 6-5 | Unit cell packing of StCaiT

Packing of StCaiT within the unit cell $P6_5$ with the cell parameters $a = b = 128.8$ Å, $c = 309.5$ Å, $\alpha = \beta = 90°$, $\gamma = 120°$.

Appendix

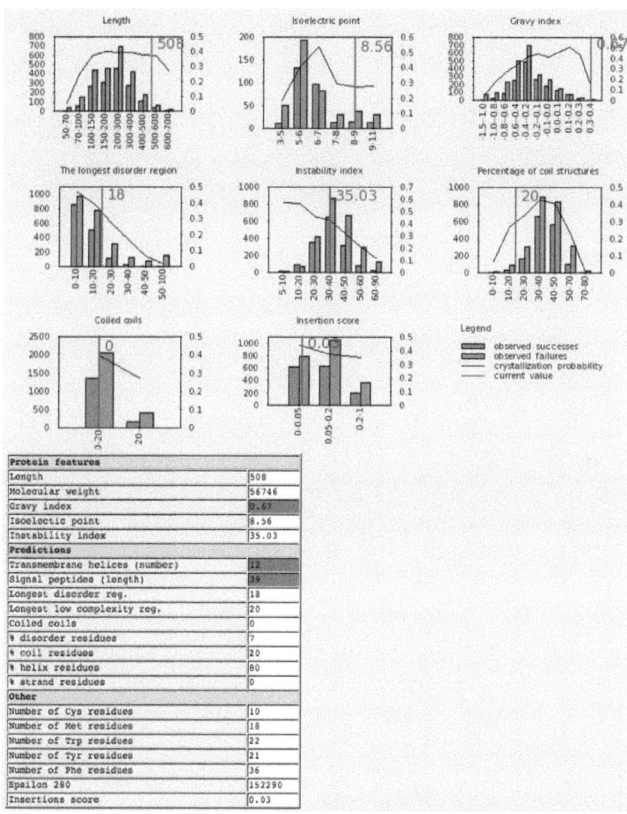

Figure 6-7 | Prediction of the crystallization probability for PmCaiT with linker peptide

When the small linker peptide connecting the amino terminus to the His$_6$ tag in PmCaiT is included in the calculations for the crystallization feasibility prediction the hydrophobicity index decreases to 0.67 and the pI value increases to 8.56, which is still located in the "observed success" region.

Appendix

7 Acknowledgements

Ich danke Herrn Prof. Werner Kühlbrandt für das Vertrauen und die Freiheit in der Projektplanung und –durchführung. Weiterhin danke ich ihm für die wertvolle Unterstützung als Doktorvater und Mentor. Es war eine sehr lehrreiche Zeit und ich bin sehr dankbar, dass ich meine Doktorarbeit in seiner Abteilung durchführen durfte.

Herrn Prof. Volker Dötsch danke ich für die Übernahme der Betreuung an der Johann Wolfgang Goethe – Universität und für die Erstellung des Gutachtens.

I would like to thank Dr. Anke Terwisscha van Scheltinga for being my supervisor in X-ray crystallography. I very much enjoyed the time we had worked together at the synchrotron and I am deeply gratefull that you shared your knowledge in X-ray crystallography with me.

Stefan Köster ist ein ausgezeichneter Wissenschaftler, zuverlässiger Kollege und guter Freund. Ich bin ihm sehr dankbar für seine Hilfe im CaiT-Projekt, für die motivierenden Worte und für sein ansteckendes Lachen.

Ulrike Geldmacher danke ich für ihre Unterstützung im Labor und die hervorragende Zusammenarbeit vor allem in der letzten Phase der Doktorarbeit.

Dr. Christine Ziegler danke ich, dass sie mich in ihrem Labor aufgenommen hat. Ich danke ihr für die wissenschaftlichen Diskussionen und Ratschläge. Darüber hinaus bin ich ihr sehr dankbar für das kritische Korrekturlesen meiner Doktorarbeit.

Ich danke Dr. Özkan Yildiz für die Fahr- und Betreuungsbereitschaft bei den Synchrotrontrips und die Hilfe bei kristallographischen Fragen sowie bei Computer-Problemen. Vielen Dank, dass du die X-ray Programme immer auf dem aktuellsten Stand hälst.

Acknowledgements

Bedanken möchte ich mich auch bei den ehemaligen und aktuellen Doktoranden im Labor L2.090, die für eine harmonische Arbeitsatmosphäre sorgten. Vielen Dank Dr. Susanne Ressl, Dr. Sonja Kuhlmann, Eva Schweikhard, Katrin Rohde, Camilo Perez und Rebecca Gärtner. Eva Schweikhard möchte ich darüber hinaus für die vielen gemeinsam überwundenen Laufkilometer danken.

Jonna Hakulinen danke ich für die viele wertvollen wissenschaftlichen und privaten Diskussionen.

I would like to thank Dr. Tiago Barros and Remco Wouts for giving me helpful tips in using my Mac more efficiently.

I am thankful to Dr. Karen Davies, who read parts of this thesis and corrected my "germanized English".

Besonders danken möchte ich auch Monika Hobrack. Sie ist jederzeit sehr hilfsbereit und hat mir bei administrativen Angelegenheiten so manches Mal das Leben erheblich erleichtert.

Für mehr oder weniger wissenschaftliche Gespräche zwischendurch danke ich Dr. Tiago Barros, Dr. Fuensanta Martinez-Ruboco, Dr. Panchali Goswami, Dr. Dilem Hizlan, Thorsten Althoff, Friederike Joos und Dr. Karen Davies.

An dieser Stelle möchte ich mich ganz besonders bei meinen Eltern für ihre Liebe, ihr Vertrauen in mich und ihre ständige Unterstützung bedanken. Weiterhin danke ich meinem Bruder Mario dafür, dass er immer für sein kleines Schwesterchen da ist wenn er gebraucht wird.

Beatrice Buchin war eine der beeindruckensten Menschen, die ich in meinem Leben bisher kennengelernt habe. Sie war eine scharfsinnige, junge Wissenschaftlerin und eine sehr gute Freundin. Beatrice ermutigte mich bei dem Vorhaben an Membranproteinen zu arbeiten und meine Doktorarbeit am MPI für Biophysik in

Acknowledgements

Frankfurt durchzuführen. Trotz ihres Lebenswillens war ihr selbst die Zeit leider nicht mehr gegeben um eine Doktorarbeit zu schreiben, daher widme ich Beatrice meine Doktorarbeit.

Curriculum Vitae

Publications

Schulze S., Köster S., Geldmacher U., Terwisscha van Scheltinga A.C., Kühlbrandt W. (2010) Structural basis of Na^+-independent and cooperative substrate/product antiport in CaiT. *Nature, in press*

Hennig S., Strauss H.M., Vanselow K., Yildiz Ö., **Schulze S.**, Arens J., Kramer A., Wolf E. (2009) Structural and functional analyses of PAS domain interaction of the clock proteins *Drosophila* PERIOD and *mouse* PERIOD2. *PLoS Biol.*, Apr 28; 7 (4): e94

Yildiz Ö., Doi M., Yujnovsky I., Cardone L., Berndt A., Hennig S., **Schulze S.**, Urbanke C., Sassone-Corsi P., Wolf E., (2005) Crystal structure and interactions of the PAS repeat region of the *Drosophila* clock protein PERIOD. *Mol. Cell*, Jan 7; 17 (1): 69-82

I want morebooks!

Buy your books fast and straightforward online - at one of world's fastest growing online book stores! Environmentally sound due to Print-on-Demand technologies.

Buy your books online at
www.morebooks.shop

Kaufen Sie Ihre Bücher schnell und unkompliziert online – auf einer der am schnellsten wachsenden Buchhandelsplattformen weltweit! Dank Print-On-Demand umwelt- und ressourcenschonend produziert.

Bücher schneller online kaufen
www.morebooks.shop

KS OmniScriptum Publishing
Brivibas gatve 197
LV-1039 Riga, Latvia
Telefax: +371 686 204 55

info@omniscriptum.com
www.omniscriptum.com

Printed by Books on Demand GmbH, Norderstedt / Germany